VOLUME THIRTY SIX

THE ENZYMES

Natural Products and Cancer Signaling: Isoprenoids, Polyphenols and Flavonoids

VOLUME THIRTY SIX

THE ENZYMES

Natural Products and Cancer Signaling: Isoprenoids, Polyphenols and Flavonoids

Edited by

S. ZAHRA BATHAIE

Department of Clinical Biochemistry,
Tarbiat Modares University, Tehran, Iran, and
Department of Microbiology,
Immunology and Molecular Genetics,
Jonsson Comprehensive Cancer Center,
Molecular Biology Institute, University of California,
Los Angeles, California, USA

FUYUHIKO TAMANOI

Department of Microbiology, Immunology and Molecular
Genetics (MIMG), University of California,
Los Angeles (UCLA), Los Angeles, California, USA

ELSEVIER

AMSTERDAM • BOSTON • HEIDELBERG • LONDON
NEW YORK • OXFORD • PARIS • SAN DIEGO
SAN FRANCISCO • SINGAPORE • SYDNEY • TOKYO
Academic Press is an imprint of Elsevier

AP

Academic Press is an imprint of Elsevier
32 Jamestown Road, London NW1 7BY, UK
525 B Street, Suite 1800, San Diego, CA 92101-4495, USA
225 Wyman Street, Waltham, MA 02451, USA
The Boulevard, Langford Lane, Kidlington, Oxford OX5 1GB, UK

First edition 2014

ISBN: 978-0-12-802215-3
ISSN: 1874-6047

For information on all Academic Press publications
visit our website at store.elsevier.com

CONTENTS

CONTRIBUTORS

Javad Alizadeh
Department of Human Anatomy and Cell Science, College of Medicine, University of Manitoba, Winnipeg, Manitoba, Canada

Sumit Arora
Department of Oncologic Sciences, Mitchell Cancer Institute, University of South Alabama, Mobile, Alabama, USA

Courey Averett
Department of Oncologic Sciences, Mitchell Cancer Institute, University of South Alabama, Mobile, Alabama, USA

S. Zahra Bathaie
Department of Clinical Biochemistry, Tarbiat Modares University, Tehran, Iran, and Department of Microbiology, Immunology and Molecular Genetics, Jonsson Comprehensive Cancer Center, Molecular Biology Institute, University of California, Los Angeles, California, USA

Arun Bhardwaj
Department of Oncologic Sciences, Mitchell Cancer Institute, University of South Alabama, Mobile, Alabama, USA

Azam Bolhassani
Department of Hepatitis and AIDS, Pasteur Institute of Iran, Tehran, Iran

Chu Chen
Institute of Pharmaceutical Research, Sichuan Academy of Chinese Medicine Sciences, Chengdu, P.R. China

Jayson X. Chen
Department of Chemical Biology, Ernest Mario School of Pharmacy, Rutgers, The State University of New Jersey, Piscataway, New Jersey, USA

Jun-Rong Du
Department of Pharmacology, West China School of Pharmacy, Sichuan University, Chengdu, P.R. China

Saeid Ghavami
Department of Human Anatomy and Cell Science, College of Medicine; Manitoba Institute of Child Health, University of Manitoba, Winnipeg, Manitoba, Canada, and Health Policy Research Centre, Shiraz University of Medical Science, Shiraz, Iran

Chin-Lin Hsu
School of Nutrition, Chung Shan Medical University, and Department of Nutrition, Chung Shan Medical University Hospital, Taichung, Taiwan

Amir Kiumarsi
Chang School of Continuing Education, Ryerson University, Toronto, Ontario, Canada

Young Sup Lee
School of Life Sciences, BK21 Plus KNU Creative BioResearch Group, Kyungpook National University, Daegu, Republic of Korea

Fang-Yi Long
Department of Pharmacology, West China School of Pharmacy, Sichuan University, Chengdu, P.R. China

Hassan Marzban
Department of Human Anatomy and Cell Science, College of Medicine, and Manitoba Institute of Child Health, University of Manitoba, Winnipeg, Manitoba, Canada

Yoichi Matsuo
Department of Gastroenterological Surgery, Nagoya City University Graduate School of Medical Sciences, Nagoya, Japan

Adel Rezaei Moghadam
Faculty of Veterinary Medicine, Tabriz Branch, Islamic Azad University, Tabriz, Iran

Siddavaram Nagini
Faculty of Science, Department of Biochemistry and Biotechnology, Annamalai University, Annamalainagar, Tamil Nadu, India

Raheem Shahzad
School of Applied Biosciences, College of Agriculture and Life Sciences, Kyungpook National University, Daegu, Republic of Korea

Adeeb Shehzad
School of Life Sciences, BK21 Plus KNU Creative BioResearch Group, Kyungpook National University, Daegu, Republic of Korea

Shahla Shojaei
Department of Biochemistry, Recombinant Protein Laboratory, Medical School, Shiraz University of Medical Sciences, Shiraz, Iran

Ajay P. Singh
Department of Oncologic Sciences, Mitchell Cancer Institute, and Department of Biochemistry and Molecular Biology, College of Medicine, University of South Alabama, Mobile, Alabama, USA

Seema Singh
Department of Oncologic Sciences, Mitchell Cancer Institute, University of South Alabama, Mobile, Alabama, USA

Hiromitsu Takeyama
Department of Gastroenterological Surgery, Nagoya City University Graduate School of Medical Sciences, Nagoya, Japan

Fuyuhiko Tamanoi
Department of Microbiology, Immunology and Molecular Genetics, Jonsson Comprehensive Cancer Center, Molecular Biology Institute, University of California, Los Angeles, California, USA

Hong Wang
Department of Chemical Biology, Ernest Mario School of Pharmacy, Rutgers, The State University of New Jersey, Piscataway, New Jersey, USA

Chung S. Yang
Department of Chemical Biology, Ernest Mario School of Pharmacy, Rutgers, The State University of New Jersey, Piscataway, New Jersey, USA, and International Joint Research Laboratory of Tea Chemistry and Health Effects, Anhui Agricultural University, Hefei, PR China

Gow-Chin Yen
Department of Food Science and Biotechnology, and Agricultural Biotechnology Center, National Chung Hsing University, Taichung, Taiwan

Jinsong Zhang
International Joint Research Laboratory of Tea Chemistry and Health Effects, Anhui Agricultural University, Hefei, PR China

Haseeb Zubair
Department of Oncologic Sciences, Mitchell Cancer Institute, University of South Alabama, Mobile, Alabama, USA

PREFACE

Anticancer activities of compounds from natural resources have been documented extensively in recent decades. However, molecular mechanisms of the action of these compounds need to be further elucidated. In particular, it will be important to understand the signaling pathways targeted by these natural compounds. We have realized that various recent activities have started to shed new lights into this problem. To capture these developments, we decided to put together a volume describing recent studies concerning the role of natural compounds in cancer therapy and cancer prevention. We believe that compiling the knowledge on elucidating targets of natural compounds is important, as it may provide hints about future developments such as possible combination therapies.

In this volume, we described studies on isoprenoids, polyphenols, and flavonoids. In future volumes, we plan to cover other classes of natural products.

We are very grateful to the authors for their effort in providing excellent and informative chapters in a timely fashion. We also thank Mary Ann Zimmerman and Helene Kabes of Elsevier for their guidance and encouragement during the preparation of this volume.

S. Zahra Bathaie
Fuyuhiko Tamanoi
UCLA, Los Angeles
September 2014

CHAPTER ONE

Introduction

S. Zahra Bathaie*,†, Fuyuhiko Tamanoi†,1
*Department of Clinical Biochemistry, Tarbiat Modares University, Tehran, Iran
†Department of Microbiology, Immunology and Molecular Genetics, Jonsson Comprehensive Cancer Center, Molecular Biology Institute, University of California, Los Angeles, California, USA
[1]Corresponding author: e-mail address: fuyut@microbio.ucla.edu

Contents

Abstract

Natural products and phytochemicals have extensively attracted for their various biological effects, especially for both treatment and prevention of cancer. In this book, we try to introduce various phytochemicals as cancer therapy targets with emphasize on their effect on signal transduction pathways and their molecular targets.

Natural products are chemical substances produced by living organisms and have distinctive biological and pharmacological effects, even if they can be prepared by total synthesis. These are foreign to humans (i.e., xenobiotics) and are subject to the same pharmacological issues encountered by synthetic therapeutic agents [1].

Plants and microbes are two important sources of natural products. Antibiotics are the oldest biologically active compounds separated from microbes and used as drug to cure various human diseases, especially cancer.

Among the wide range of biologically active compounds obtained from different sources in nature, medicinal uses of plants possibly are the oldest one and came back to the ancient time, as they used by various nations. Historically, various parts of the plants, such as fruit, flower, leaves, stalks, root, seed, and even the whole plant, have been used as the home remedy. Different methods have also been used for preparation of the herbal remedy; they include preparation of the pills, capsules, or sachets from the powder; decoctions (boiled); infusion; extraction with water or oil; and so on [2]. All of these preparations have been used orally. However, some preparations may have the topical application. Thus, they can be used for inhalation therapy or in a mixture as the skin cream. In all of the above-mentioned

The Enzymes, Volume 36
ISSN 1874-6047
http://dx.doi.org/10.1016/B978-0-12-802215-3.00001-X

methods, the main goal is the efficient absorption of the effective ingredient(s) in the body.

Nowadays, targeted therapy or molecularly targeted therapy, especially in cancer treatment, has been considered to accomplish more effective treatment with less harmful effect to normal cells. In this regard, scientists try to use the more effective ingredient of the herbals, instead of the crude extract to achieve the more powerful therapy with no or minimum side effect. Therefore, fractionation, purification, and characterization of the active components have been extensively considered.

Phytochemicals (from the Greek word *phyto*, meaning plant) are compounds found in plants. They are biologically active and provide health benefits for humans. These chemicals have metabolic or protective role in their own plants, but may exert the same or other effects in other organisms like animal or human bodies.

Table 1.1 shows the overall classification of phytochemicals according to their chemical structure, biological activity, and plant sources. Different chapters of this book reviewed the mechanism(s) of the anticancer effect of some of these phytochemicals.

Various biological or pharmacological activities have been reported for phytochemicals. Some of them include antimicrobial, antivirus, or antifungal effects; antioxidant activity or activation of the antioxidant defense system; modulation of the detoxifying enzymes; stimulation or suppression of the immune system; decrease of platelet aggregation; modulation of hormone metabolism; and regulation of the metabolism of the building blocks in the body.

Chemoprevention and anticancer property are two novel approaches emphasizing the prevention or delay of carcinogenesis, or treatment of cancer by means of natural products through pharmacologic, biologic, and even nutritional intervention. This involves the discovery and characterization of the phytochemicals as a new drug with specific effect on cell cycle proteins, growth factors, or hormone receptors, and/or specific inhibitory or activatory effect on specific enzymes. Chemotherapeutic and chemoprevention by targeting key components of the apoptosis pathways, cell cycle checkpoints, autophagy regulation, ER stress response, and protein folding targets are the main goal of the new drug design approaches (Fig. 1.1).

Targeting of the tumor microenvironment, more particularly inflammatory mediators and reactive oxygen/nitrogen species; upregulation of intercellular communication through gap junction or tight junction; regulation of upstream kinases of intracellular signaling cascades or downstream transcription factors; elimination of endogenous and environmental

Table 1.1 Family and Chemical Structure of Phytochemicals Found in Plants

Row	Family	Chemical Structure	Component	Plant Name
1	Isoprenoids or terpenoids	Monoterpenoids	Zingerone or vanillylacetone	Ginger
2			Terpineol	
3			Picrocrocin	Saffron
4		Diterpenoids	Taxol (or paclitaxel)	*Taxus brevifolia*
5		Triterpenoids	Saponins	Many plants
			Ganoderic acid	*Ganoderma* mushrooms
			Lucidenic acid	*Ganoderma* mushrooms
			Ursolic acid	Apple, basil, bilberries
6		Tetraterpenoids or carotenoids	Beta-carotene	Carrot
7			Lycopene	Tomato
			Lutein	Spinach, kale, and yellow carrot
			Crocin and crocetin	Saffron
8		Tetranortriterpenoids	Limonoids	Citrus fruits
9		Steroids (phytosterols)	Withanolide	Tomatillo
10	Phenolics and polyphenols	Curcuminoid	Curcumin	Turmeric
11		Stilbenoid	Resveratrol	Skin of red grapes
12		Lignan	Honokiol	Magnolia
13		Chalconoids	Chalcones Geranyl chalcone Geranylgeraniol	
14		Tannins		Pomegranates, persimmon, berries, nuts

Continued

Table 1.1 Family and Chemical Structure of Phytochemicals Found in Plants—cont'd

Row	Family	Chemical Structure	Component	Plant Name
15	Flavonoid-polyphenolics	Quercetin		Radish leaves, dill, red anion
16		Kaempferol		Tea, broccoli, grapefruit
17		Flavan-3-ols	Catechins	Green tea
18	Flavonoids	Flavones	Apigenin, chrysin, luteolin	Parsley, celery, and citrus peels
		Flavanones	Eriodictiol, hesperitin	Citrus fruits
19		Isoflavonoids	Genistein	Fava beans, soybeans
20		Licoflavanone		*Glycyrrhiza glabra*
21	Anthocyanidins	Cyanidin		Grapes, bilberry, blackberry, blueberry, cherry
22	Polycyclic compounds	Quinoline alkaloid	Camptothecin	Camptotheca, happy tree
23	Aromatic compounds	Aromatic acids (hydroxycinnamates)	Cinnamic acid	Cinnamon
24			Caffeic acid	Basil, apple
25			Coumarin	Citrus fruits
26		Aromatic aldehyde	Safranal	Saffron
27	Glucosinolates	Isothiocyanates	Allyl isothiocyanate	Mustard, radish, horseradish
28	Vanilloids	Phenolic aldehyde	Vanillin and derivatives (vanillic acid, vanillyl alcohol, etc.)	Vanilla bean
			Capsaicin	Chili peppers
29	Organosulfurs	Allylic sulfurs	Allicin	Garlic
30			Diallyl disulfide	
31	Thiosulfates			Leek
32	Aromatic heterocyclic	Indole alkaloids Ergot alkaloids		Calabar bean seed, rye and related cereals
		Monoterpenoid alkaloids		

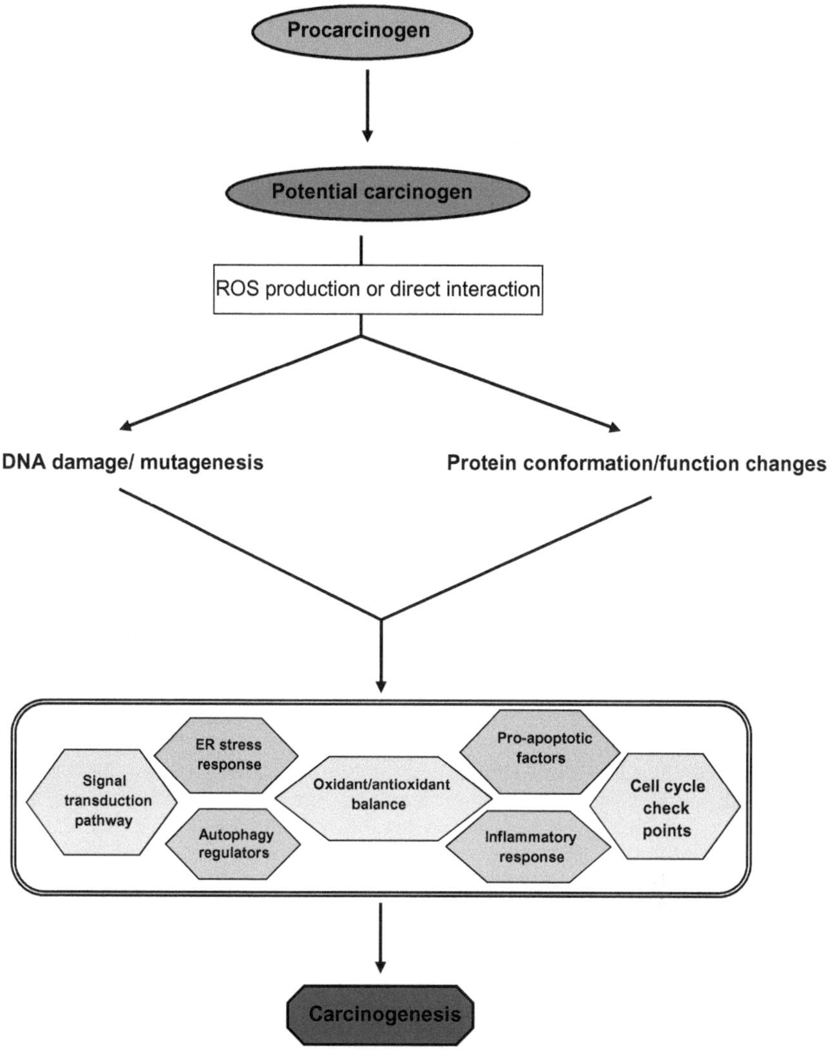

Figure 1.1 The process of carcinogenesis from beginning. Various phytochemicals and drugs can inhibit any step(s), which is extensively discussed in the chapters of this book.

carcinogens; and/or reduction of angiogenesis are some proposed chemopreventive strategies by means of pharmacological or nutritional factors.

Among the mechanisms mentioned in Fig. 1.1, reactive oxygen species (ROS) production (the cellular distortion in the balance of oxidant/antioxidant) is the most important in both carcinogenesis and killing of cancer cells. A set of 13 p53-induced genes (*PIG* genes) have a key role in the reactions resulting in the synthesis of H_2O_2, OH^-, or O_2^- superoxides. P53 induces apoptosis by turning on the synthesis of genes whose primary function is the synthesis of ROS [3,4]. In addition, not only the ionizing radiation but also the effective anticancer drugs and frame-shifting mutagens induce apoptosis through ROS production [3,5–7]. Cancer cells containing the higher levels of antioxidants and antioxidant enzymes are most difficult to treat [3]. Watson mentions that "free-radical-destroying antioxidative nutritional supplements may have caused more cancer than they have prevented" [3].

However, most natural products are a potential source of antioxidants [8]. They are not only useful for cancer prevention but also for cancer treatment. The mechanistic studies also showed their target genes in various types of cancer. Therefore, one of the main goals of this book is to discuss issues about the role of phytochemicals as the natural antioxidants in cancer treatment.

REFERENCES

[1] S. Barnes, J. Prasain, Current progress in the use of traditional medicines and nutraceuticals, Curr. Opin. Plant Biol. 8 (3) (2005) 324–328.
[2] Avicenna, The Canon of Medicine, vol. 1, Soroush, Tehran, 1997, pp. 10–43.
[3] J. Watson, Oxidants, antioxidants and the current incurability of metastatic cancers, Open Biol. 3 (120144) (2013) 1–9.
[4] K. Polyak, et al., A model for p53-induced apoptosis, Nature 389 (6648) (1997) 300–305.
[5] D. Trachootham, J. Alexandre, P. Huang, Targeting cancer cells by ROS-mediated mechanisms: A radical therapeutic approach? Nat. Rev. Drug Discov. 8 (7) (2009) 579–591.
[6] J.C. Yang, et al., Selective targeting of breast cancer cells through ROS-mediated mechanisms potentiates the lethality of paclitaxel by a novel diterpene, gelomulide K, Free Radic. Biol. Med. 51 (3) (2011) 641–657.
[7] J. Wang, J. Yi, Cancer cell killing via ROS: To increase or decrease, that is the question, Cancer Biol. Ther. 7 (12) (2008) 1875–1884.
[8] D.M. Maestri, et al., Natural Products as Antioxidants, in: Filippo Imperato (Ed.), Phytochemistry: Advances in Research, Research Signpost, Kerala, India, 2006, pp. 105–135.

CHAPTER TWO

Perillyl Alcohol (Monoterpene Alcohol), Limonene

Shahla Shojaei*[,1], Amir Kiumarsi[†,1], Adel Rezaei Moghadam[‡], Javad Alizadeh[§], Hassan Marzban[§,¶], Saeid Ghavami[§,¶,||,2]

*Department of Biochemistry, Recombinant Protein Laboratory, Medical School, Shiraz University of Medical Sciences, Shiraz, Iran
[†]Chang School of Continuing Education, Ryerson University, Toronto, Ontario, Canada
[‡]Faculty of Veterinary Medicine, Tabriz Branch, Islamic Azad University, Tabriz, Iran
[§]Department of Human Anatomy and Cell Science, College of Medicine, University of Manitoba, Winnipeg, Manitoba, Canada
[¶]Manitoba Institute of Child Health, University of Manitoba, Winnipeg, Manitoba, Canada
[||]Health Policy Research Centre, Shiraz University of Medical Science, Shiraz, Iran
[1]Both authors share equal first authorship.
[2]Corresponding author: e-mail address: saeid.ghavami@med.umanitoba.ca; saeid.ghavami@gmail.com

Contents

Abstract

Natural products have a long history of use in traditional medicines and their activities against different diseases have been the focus of many basic and clinical researches in past few decades. The essential oils, volatile liquid containing aroma compound from plants, are known as active ingredients in the herbal medicine. Perillyl alcohol (POH) is usually available through dietary sources and is being explored for its cancer chemoprevention, tumor growth suppression, and regression. Citrus peels are the waste product of juice manufacturing industries and have been considered as a critical problem for environmental green ecology policies for years. One of the most well-known approaches to overcome this problem is transformation of these

The Enzymes, Volume 36
ISSN 1874-6047
http://dx.doi.org/10.1016/B978-0-12-802215-3.00002-1

monoterpene by the use of specific strains of bacteria or yeasts. Limonene (1-methyl-4-isopropyl-cyclohexene) is a monoterpene, as other monoterpenes consists of two isoprene units, that comprises more than 90% of citrus essential oil and it exists in many fruits and vegetables. Although, the anticancer activity of D-limonene has identified nearly two decades ago, it has recently attracted much more attention in translational medicine. In this chapter, we will overview the anticancer effects of POH and D-limonene. Later, we will address the pharmacokinetics of these compounds, highlight the signaling pathways which are targeted by these proteins, review the clinical trials which have been done for these compounds in different cancer models, and finally discuss the future directions of the research in this field that might be more applicable in future cancer therapy strategies.

ABBREVIATIONS

AP-1 activator protein-1
BAK Bcl-2 homologous antagonist/killer
BAX Bcl-2-associated X protein
Bcl-2 B-cell lymphoma-2
eIF4E eukaryotic translation initiation factor 4E
ERK extracellular signal-regulated kinase
FTase farnesyl transferase
GSK glycogen synthase kinase
hTERT human telomerase reverse transcriptase
IGF insulin-like growth factor
M6P mannose-6-phosphate
M6P/IGF mannose-6-phosphate/insulin-like growth factor
Mek mitogen/extracellular signal-regulated kinase
mTOR mechanistic target of rapamycin
PARP poly ADP ribose polymerase
PI3K phosphatidylinositol-4,5-bisphosphate 3-kinase
POH perillyl alcohol
Ras rat sarcoma viral oncogene homolog
RhoB ras homolog gene family, member B
ROS reactive oxygen species
TGF-β transforming growth factor β
VEGF vascular endothelial growth factor
VEGFR1 vascular endothelial growth factor receptor-1

1. INTRODUCTION

L(−)-Perillyl alcohol (POH) also called (*S*)-(−)-POH (CAS Number 18457-55-1, Fig. 2.1) is a hydroxylated monoterpene and includes

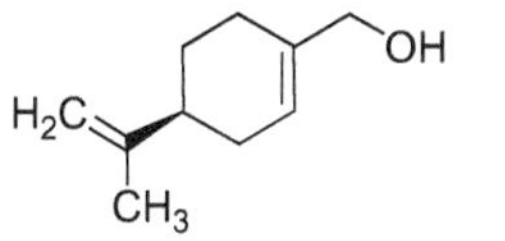

Figure 2.1 Perillyl alcohol ($C_{10}H_{16}O$).

two isoprene subunit which is metabolized by mevalonate cascade [1]. Trace amount of POH can be detected in many kinds of essential oils including lavandin, peppermint, spearmint, sage, cherries, cranberries, perilla (*Perilla frutescens*), lemongrass, wild bergamot, gingergrass, savin, caraway, and celery seeds [2,3]. It is liquid with the *boiling point* 119–121 °C/11 mmHg(l). POH has been shown to be implicated in inhibition of different stages and different types of tumor such as skin, liver, glioma, breast, lung, mammary, colon, pancreatic, gastric, and prostatic cancers in rodent models [4–7]. It also plays roles in broad spectrum of pathophysiologic processes like inflammation, oxidative stress, ornithine decarboxylase activity, thymidine incorporation into DNA, rat sarcoma viral oncogene homolog (Ras) protein family-signaling pathway, and alteration of the B-cell lymphoma-2 (Bcl-2) and Bcl-2-associated X protein (BAX) expression [4,8]. POH affects multiple different steps in the carcinogenesis process [9,4] and is a metabolite product of limonene.

Limonene is a colorless liquid hydrocarbon classified as a cyclic terpene. It is the most common terpene, making up to 95% in some citrus [10]. Lemon contains considerable amount of limonene, which is responsible for its odor. Limonene is a chiral molecule with two possible isomers, D- and L-limonene, and biological sources produce only one enantiomer. The more common isomer is the (*R*)-enantiomer, D-(+)-limonene (CAS# 5989-27-5), which possess a strong smell of orange [11,12]. The *Chemical Abstract Name is* (*R*)-1-methyl-4-(1-methylethenyl)cyclohexene, and the *IUPAC Systematic Name is* (*R*)-(+)-*para*-Mentha-1,8-diene. The racemic limonene is known as dipentene [13–15]. Limonene is liquid with the *boiling point* 175.5–176 °C [16] and *melting point* –74.3 °C [17]. It is a relatively stable terpene but will be decomposed to isoprene at elevated temperatures [18]. It can also be oxidized to carveol, carvone, and limonene

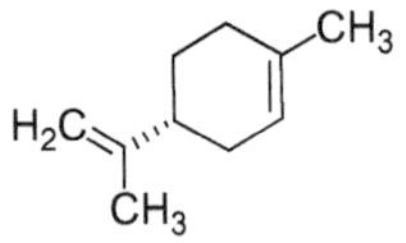

Limonene

Figure 2.2 D-Limonene ($C_{10}H_{16}$).

oxide [19]. It is slightly soluble in water (13.8 mg/mL at 25 °C) and soluble in acetone, dimethyl sulfoxide, and ethanol [20]. The structural formula of *R*- and *S*-limonene enantiomers is shown in Fig. 2.2.

Limonene does not have any functional group available for hydrolysis; its cyclohexene ring and ethylene group are chemically resistant to hydrolysis. Biological degradation has been observed in some species of microorganisms, such as *Penicillium digitatum, Corynespora cassiicola, Diplodia gossypina,* and a soil strain of *Pseudomonas sp.* [21,22].

D-Limonene is obtained commercially from citrus fruits through two primary methods: centrifugal separation or steam distillation. It is mainly produced in Australia, Brazil, Germany, Japan, and the United States [23].

The main industrial use of limonene is as a precursor to carvone or α-terpineol [24]. Other uses of limonene are as a fragrance in cosmetics and food products, as a component in industrial solvents and aromatherapy [25]. The extraction method differs depending on the final application. Thus, for pharmaceutical and food uses, the preferred extraction methods are steam distillation and cold expression. For use in perfumes, other methods such as extraction with lipophilic solvents or supercritical fluids are used [11,25].

D-Limonene has a wide variety of applications and has been used in food industries as flavors and in chemical industries as solvent and resins [14,16]. In contrast, L-limonene has a piney, turpentine-like odor [26]. It has shown encouraging and well-established chemopreventive activity against many types of cancers [27,28]. In the following sections, we will discuss different aspects of anticancer activity of these compounds.

The principle metabolites of limonene are (+)- and (−)-*trans*–carveol, (+)- and (−)-POH, perillic acid, iso-pipiritenol, α-terpineol, limonene-1,2-epoxide, limonene-1,2-diol, and limonene-8,9-epoxide—a product of 7-hydroxylation by CYP2C9 and CYP2C19 cytochromes in human liver microsomes [29]. Other products obtained biologically from limonene are carvone, perillaldehyde. The metabolic pathway of D-limonene has been shown in Fig. 2.3 [12,15,30,31].

In the following sections, we will explain different mechanisms, which are involved in POH and limonene anticancer effects, their pharmacokinetics, and different trials that they have been used.

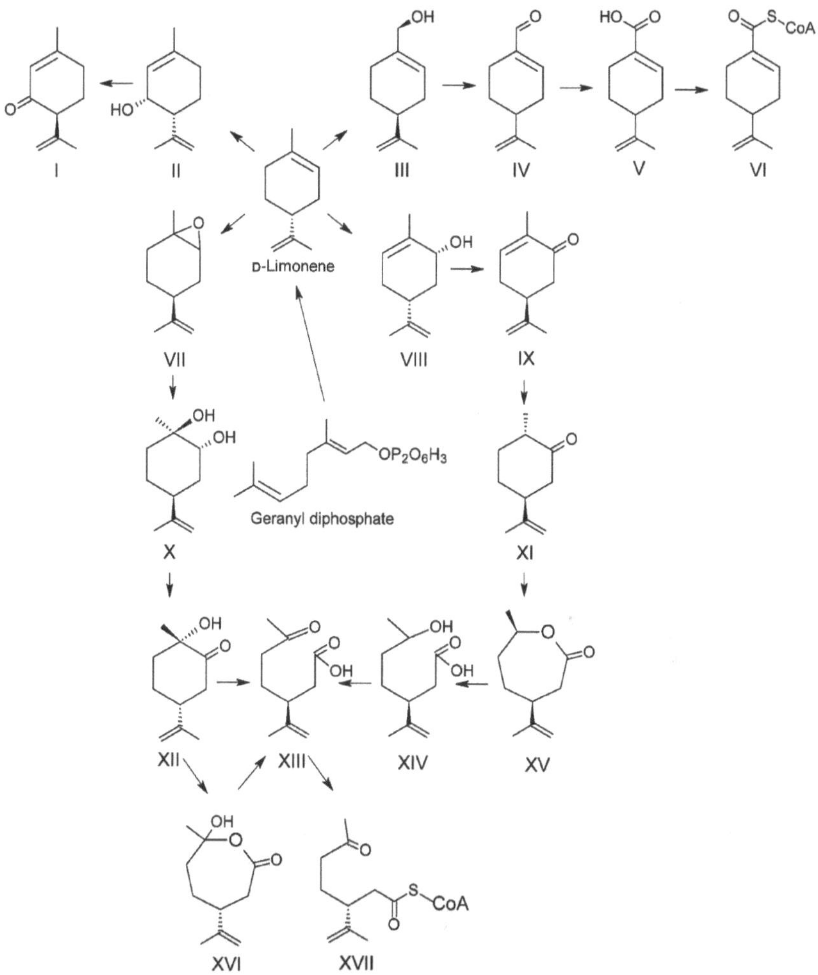

Figure 2.3 Biotransformation pathway of D-limonene and geranyl-diphosphate (GPP) in plants or microorganisms. I: Isopiperitenone; II: isopiperitenol; III: perillyl alcohol; IV: perillyl aldehyde; V: perillyl acid; VI: perillyl-CoA; VII: limonene-1,2-epoxide; VIII: carveol; IX: carvone; X: limonene-1,2-diol; XI: dihydrocarvone; XII: 1-hydroxy-2-oxolimonene; XIII: 3-isopropenyl-6-oxoheptanoate; XIV: 6-hydroxy-3-isopropenylheptanoate; XV: 4-isopropenyl-7-methyloxepan-2-one; XVI: (4R)-7-hydroxy-4-isopropenyl-7-methyloxepan-2-one; XVII: 3-isopropenyl-6-oxoheptanoyl-CoA. *Data from Refs. [30,12,31,15].*

2. PERILLYL ALCOHOL

2.1. Perillyl Alcohol Mechanism of Action in Cancer Therapy and Pharmacokinetics

Although it is not clear exactly how POH elicits its antitumor effects, a number of potentially important mechanisms have been reported. One potential mechanism is the induction and augmentation in expression of the transforming growth factor-beta (TGF-β) signal transduction pathway and then initiation of apoptosis followed by induction of cytostasis in tumor cells with no impact on normal cells [32]. There are, however, other possible mechanisms including, for example, inhibition of downstream Ras-signaling pathways, modulation of AP-1 (activator protein-1) activity, early G1 arrest, differentiation, inhibiting the isoprenylation of small Rho-GTPase proteins, induction of growth factors, and modulating the activity of cell cycle checkpoint proteins [33–36] (Fig. 2.4).

POH can act as a regulator of cholesterol biosynthesis by inhibiting of mevalonate pathway [37]. In the line with the effect on cholesterol biosynthesis, it has been reported that almost 10% decreased body weights in rats treated with POH were observed in a study, in which it was attributed to a decrease in body fat [35]. Prior studies also showed that POH was able to affect mevalonate pathway by blocking the conversion of 3-hydroxy-3-methylglutaryl-coenzyme A to mevalonic acid followed by suppressing the synthesis of small Rho-GTPase proteins and 3-hydroxy-3-methylglutaryl-coenzyme A reductase, leading to decrease the activity of isoprenylated Ras and Ras-related proteins [32,38]. Thereby, it is considered as a prenyltransferase inhibitor [38]. Farnesylation is necessary for mutant Ras activity, and farnesyl transferase (FTase) inhibitors block the oncogenic activity of Ras [39]. It is, however, an interesting claim that this monoterpene can have anticarcinoma effect on leukemia cells in a Ras-independent manner [40]. This finding supported the idea that there is other possible mechanism(s) leading to its cancer chemopreventive effect on different types of malignant cells in both a Ras-dependent and a Ras-independent pathway. The inhibitory proliferation effects of POH toward pancreatic cancers were evidenced by initiation of Bcl-2 homologous antagonist/killer (BAK)-induced apoptosis a proapoptotic protein in cancer without affecting normal controls [41]. The mechanism of POH effects on liver tumors involves an increase in the rate of tumor cell loss through stimulation of TGF-β and M6P/IGF II (mannose-6-phosphate/insulin-like

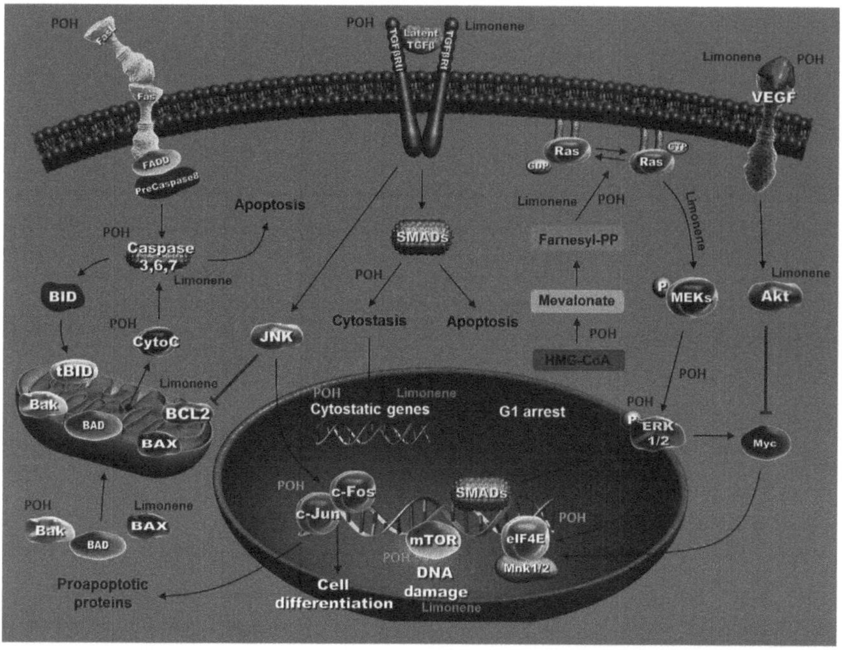

Figure 2.4 Schematic representation of antitumor effects of limonene and perillyl alcohol (POH). Limonene and POH have various inhibitory and stimulatory roles in some key pathways involved in tumor progression and regression. Both of these natural products play an important role in regulation of cell death. The limonene exerts its effects by upregulation of BAX, cytochrome *c* release, caspase-3, -9, TGF-β and downregulation of antiapoptotic Bcl-2. POH also upregulates Bak, caspase-3, FasL, TGF-β, c-fos, and c-Jun and blocks ERK1/2 phosphorylation and Mek–Erk pathway. Both of limonene and POH can inhibit tumor progression through downregulation of basal production of VEGF in cancer cells. They also suppress mevalonate pathway as well as isoprenylation of small G proteins, leading tumor regression. (See the color plate.)

growth factor-II) receptors [35]. Inhibition of mitogen/extracellular signal-regulated kinase (Mek)–extracellular signal-regulated kinase (Erk) pathway has been offered as another POH mechanism of action [42]. In agreement with this concept, another study has reported that POH play pivotal role in induction of apoptosis by blocking both ERK1/2 phosphorylation and small GTPase signaling in K562 cells. These results suggested that POH may have several targets when activated by this pathway [43] that might be critical for cancer therapy. It was also proved that POH can downregulate the basal production of vascular endothelial growth factor (VEGF) in cancer cells and upregulate the release of angiopoietin-2 from endothelial cells, causing tumor regression [43]. The targeting pathways by POH have been summarized in Fig. 2.4.

A study presents evidence that, in glioma cells, POH exerts its impact by increasing the expression of Fas Ligand which subsequently results in enhancement of chemotherapeutic delivery and efficacy [44]. The chemopreventive effect of POH has been demonstrated through inhibition of AP-1 activation induced by UVB in skin tumor promotion *in vitro* and *in vivo* [33]. This finding is in discrepancy with a study in human breast tumor cell line, T47D-C4-2W, which showed that POH-induced c-fos and c-Jun (two oncoprotein which combine each other and form AP-1) transcriptional activity via the c-Jun N-terminal kinase/stress-activated protein kinase pathway, and subsequent AP-1 activation [45]. Another preclinical study proved the involvement of POH in activation of expression of genes downstream of TGF-β, and AP-1 induction by altering the function of c-Jun and c-fos along with induction of proapoptotic genes expression like BAX and bad in mammary carcinomas [32]. Both c-Jun and c-fos are activated before and during apoptosis, and their stimulation by POH might be an early key event in the signaling cascades lead to cell death [46–49]. The suppression of androgen receptor (AR) gene expression and stimulation of the expression of c-Jun by POH attenuate AR-mediated action in androgen-responsive prostate cells [50]. POH has also been shown to be involved in inhibition of cell growth, cell cycle progression, and cyclin D1 gene expression in human breast cancer cell lines [51]. Other *in vitro* studies reported an inhibitory effect of POH in the growth of pancreatic and lung cancer cells [52]. It is established that the decrease in cyclin D1 and E followed by enhanced p21 (Cip1/Waf1) and decreased PCNA expression contributes to the cell cycle arrest in human breast cancer cells in culture [53]. These are the mechanism of antiproliferative and antimetastatic function of POH against human breast cancer cells in the nude mice system [53]. The impact of POH in glial C6 cell line and chick embryo chorioallantoic membrane model has been investigated by Balassiano *et al.*, who showed that it played a function as an antimetastatic molecule in both *in vitro* and *in vivo* [54]. An *in vitro* experiment suggested that POH suppressed breast cell migration without affecting cell adhesion by disrupting the cytoskeletal machinery required to exert the necessary force for lamellar extension [55]. *In vitro* studies indicated that POH was responsible for induction of differentiation in the neuroblastoma-derived cell line, Neuro-2A cells [36]. Incubation of these cells with POH showed that it suppressed DNA synthesis [36]. The histologic evaluation of the sites that induced regression of tumor growth by the application of POH showed that it is implicated in redifferentiation of the tumor cells [3]. Recently, an additional mechanism has proposed for repression of prostate cancer progression by POH that can regulate telomerase activity via unique synergistic decreases of hTERT (human telomerase reverse transcriptase) protein translation and disruption of the hTERT–mTOR (mechanistic target of rapamycin)–RAPTOR protein complex [56]. The results confirmed the hypothesis of oncogenic shock [57]. They found that the regulation of telomerase and TERT protein by POH is associated with high eIF4E (eukaryotic translation initiation factor 4E) levels, and have no effect in normal pMV7 cells with low eIF4E levels [57]. A summary of POH mechanism of action has been summarized in Fig. 2.4.

2.2. Perillyl Alcohol Biosafety and Adverse Effects in Clinical Application and Clinical Trials

In 1995, a Phase I dose-escalation study in dog was performed to investigate POH biosafety, pharmacokinetics, and antitumor activity. The most common side effect of POH in this experiment was observed in gastrointestinal (nausea, early satiety, eructation, and unpleasant taste) and fatigue [58]. Then, preclinical toxicity and pharmacology studies in rats and dogs were tested by National Cancer Institute (NCI) and showed that administration of POH affects the renal and GI functions, which high single dose caused emesis and diarrhea [59]. They also found forestomach epithelial hyperplasia, necrosis, and renal tubular degeneration in histopathology evaluation [59]. Later, mild to moderate toxicity was observed in Phase I trials on daily orally administration [60–62].

In human Phase I POH study, gastrointestinal tract side effects were reported as some gastrointestinal symptoms including nausea, vomiting, and diarrhea [62–65]. The data obtained from Phase II clinical trials of orally administered POH in advanced ovarian, metastatic colorectal, and metastatic breast cancers were similar to those obtained results on the dose-escalation study and Phase I testing [64,66,67]. Increase in side effects has been reported after extremely high doses usage to induce POH systemic activity [68]. Significant heterogeneity in tolerance and pharmacokinetics of POH was found in different patients [61]. Tumor shrinkage and prolonged survival were observed in clinical trials in various malignancies [62]. Phase I trial by Hudes *et al.* in 17 patients administrated with POH on a three times daily dosing schedule for 14 consecutive days was performed and the maximum tolerated dose was found 6300 $mg/m^2/day$ but some side effect symptoms were observed in this dosage, including chronic nausea and fatigue. The grade 1–2 hypokalemia as a common side effect were reported and one patient developed grade 3 mucositis at the 8400 mg/m^2 per day [62]. Several groups have been measured POH and its metabolite levels in serum, plasma, and urine using gas chromatographic-mass spectrometric analysis. The findings from these investigation showed that there were no correlation between toxicity and patient tolerance with maximum concentrations (C_{max}) and estimated area under the curve [63]. A Phase I study of topical POH cream formulation has been tested and showed acceptable skin tolerance [69]. Another Phase I/II study has displayed no toxicity or pharmacokinetic interaction when POH intranasal was administered in a four times daily schedule in patients. In this work, POH intranasal administration had antitumor activity and acceptable tolerance in malignant gliomas [70]. However, a clinical trial study has presented inconsistent results with previous studies. The administration of women with POH four times daily at 1200–1500 mg m/2 dose/1 on a 28-day cycle demonstrated a lack of response and poor tolerance of this regimen, having no beneficial effects on advanced treatment-refractory breast carcinoma [67]. Some other previous clinical studies have been obtained results also revealed that POH has no effect in the treatment of solid tumors [65,66]. The summary of POH anticancer activity in different *in vitro* and *in vivo* model has been summarized in Table 2.1.

3. LIMONENE

3.1. Limonene Pharmacokinetics

Limonene is directly absorbed in gastrointestinal tract of both humans and animals when administered orally [71,72] and rapidly disperses to different tissues (detectable in serum, liver, lung, kidney, and many other tissues) and quickly undergoes through the metabolization processes for hydroxylation and carboxylation to produce more soluble metabolites like perillic acid, dehydroperillic acid, limonene-1,2-diol and limonene-8,9-diol [73,74]. POH is the precursor of perillic acid and have more potent antiproliferative effect than limonene and perillic acid. Despite low doses necessary to treat gallstone, cholecystitis, and angiocholitis [75], higher doses of limonene is required to induce anticancer activity [76,77]. With this regard, pharmacokinetic study was arranged on patients with advanced cancers and the patients showed acceptable tolerance of limonene [71] and other teams have tried to develop methods to measure concentration of this compound in the blood and other samples. For example, bioavailability of D-limonene in

Table 2.1 Summary of Preclinical Studies of Perillyl Alcohol in Cancer Therapy

Type of Cancer	Route of Administration	Type of Study	Mechanism of Action	Reference
Apoptosis pathway as the main mechanism				
Mammary carcinomas	Orally	Rat	Induction of apoptosis; upregulation of TGF-β and M6P/IGF	[32]
Pancreatic cancer	Treatment	D27-B12/13, BxPC-3; MIA PaCa-2, Panc-1 cells	Initiation of Bak	[41]
Liver tumor	Orally	Fischer 344 rats	Upregulation of TGF-β and M6P/IGF	[35]
Leukemia	Treatment	Bcr/Abl-transformed murine; FDC. P1 and 32D cell lines	Inhibition of the Mek–Erk pathway	[42]
Leukemia	Treatment	BLMVECs and K562 cells	Blocking both ERK1/2 phosphorylation and small GTPase signaling, inhibition of angiogenesis	[43]
Glioma	Treatment	T98G cell line	Increasing the expression of FasL	[44]
Breast	Treatment	T47D-C4-2W cell line	Modulation of c-fos and c-Jun activity	[45]
Other mechanisms				
Breast cancer	Treatment	T-47D, MCF-7 and MDA-MB-231 cells	Cell cycle arrest; decreased cyclin D1 gene expression	[51]
Breast cancer	Injection	Nude mice	Antiproliferative; antimetastatic activities	[53]

Continued

Table 2.1 Summary of Preclinical Studies of Perillyl Alcohol in Cancer Therapy—cont'd

Type of Cancer	Route of Administration	Type of Study	Mechanism of Action	Reference
Leukemia	Injection	Chick embryo chorioallantoic membrane model	Antimetastatic activity	[54]
Neuroblastoma	Treatment	C6 glial cell line	Inhibition cell proliferation	[54]
Breast cancer	Treatment	MCF-10A, MDA-MB 435 cell lines	Disrupting the cytoskeletal machinery; regulation of isoprenylation	[55]
Neuroblastoma	Treatment	Neuro-2A cells	Induction of differentiation; inhibition of ubiquinone (CoQ) synthesis	[36]
Prostate cancer	Treatment	DU145 cell line	Regulation of hTERT via mTOR	[56]
CHO-derived cells	Treatment	pMV7 and rb4E cell lines	Overexpression of eIF4E inhibit telomerase and TERT activity	[57]

mammary gland has been measured after its oral administration in SKH-1 mice model [78]. This study showed mammary tissue disposition of D-limonene with no clinical signs of toxicity and high D-limonene levels in mammary tissue without affecting the morphology of normal mammary gland [78].

3.2. Limonene Anticancer Activity and Clinical Trials

D-Limonene inhibits cell growth in various cancer models such as pancreatic, mammary, breast, liver, stomach, lung, prostatic tumors, skin and forestomach cancers [71,79,80], while it is nontoxic to normal cells and tissues [28,72,78,15]. Also, there are reports on antioxidant activity of D-limonene [81] and its ability to suppress the production of proinflammatory mediators [82,83]. It has been shown that D-limonene had significant potency to treat a variety of cancers in animal studies [76], including breast, liver, pancreatic carcinomas, neuroblastomas, and leukemias [84,85]. Moreover, some animal trials demonstrated D-limonene-mediated inhibition of carcinogen-induced neoplasia when administered orally [86]. In recent couple of years, D-limonene has been proved to have promising anticancer activities in both prevention and treatment of a variety of animal tumor models (at both initiation and progression phases) like mammary, breast, colon, pancreatic, gastric, and hepatic cancer [85,87–89] which was caused by chemical carcinogens [85,88]. Depending on the chemically induced medium used, inhibition occurs in either the initiation or progression phases [90]. D-limonene may have preventive [91] or therapeutic [92] effects against chemically induced rat mammary tumors and carcinomas [80]. Additionally, a case–control study proved a dose–response relationship between decreased risk of skin cell carcinoma and higher citrus (a principal source of D-limonene) consumption patterns in diet [93]. One of these studies reported that monoterpenes (including D-limonene) may have chemopreventive potential against aflatoxin-induced liver cancer and inhibited the formation of aflatoxin-DNA (AFB1-DNA) adducts in male F344 rats fed by D-limonene. In line with the results from this study, another study has shed some light on the underlying role of D-limonene preventing the formation of tumors as it has demonstrated compelling results with other cancer-inducing chemicals [94]. As a result, Phase I/II therapeutic clinical trials have been conducted in patients with advanced cancer to evaluate potential cancer chemotherapeutic activities of different D-limonene pharmacological preparations along with to elucidate its mechanism of action [71,73]. Results from these trials are very promising and shows potential for new cancer therapy and prevention strategies. A Phase I/II trial with orally administered D-limonene was done in patients with cancer that demonstrated disease stabilization in three patients with colon cancer and a partial response in one patient with breast cancer [71]. Therefore, based on these findings, it appears that these *in vivo* and *in vitro* studies have contributed to elucidate the possible role of D-limonene while highlighting its potential in cancer prevention and treatment.

3.3. Limonene Mechanisms of Action, Targets, and Clinical Applications

So far, many studies have done to probe the mechanisms of preventive and therapeutic properties of D-limonene in multiple types of cancer. Tumor angiogenesis prevention and increase in apoptosis of tumor cells are suggested as two primary anticancer mechanisms of D-limonene [89]; however, there are other possible mechanisms. Here, we present and discuss some of the well-proved D-limonene mechanisms of action, targets, and clinical applications in cancer prevention and/or therapy.

It has been demonstrated that D-limonene and its metabolites can inhibit the isoprenylation (both protein farnesylation and geranylgeranylation) [34] of a subset of proteins leading to alteration of RAS signaling [77] such as small G proteins, including p21ras farnesylation which is a component of cell growth-signaling pathway [34,95]. It implies that D-limonene inhibits the activity of FTase enzyme. The majority of these isoprenylated proteins such as p21ras affected by D-limonene have a molecular weight of 21–26 kDa [77] and are involved in regression of tumors [91]. Isoprenylation is a posttranslational modification that plays a pivotal role in the functional activity of many proteins which are involved in cell proliferation, cell growth, cell transformation, and cell growth-signaling pathways like Rho-GTPase and Ras family [96,97]. In fact, impairment of protein prenylation might also account for the antitumor activity of D-limonene [79]. Besides, the blocking of isoprenylation of small G proteins, D-limonene has also been observed to have a wide range of other cellular effects, including the inhibition of coenzyme Q synthesis [84], induction of various growth factors and their receptors [98] and induction of Phase I and Phase II carcinogen-metabolizing enzymes (cyt p450) [76]. For example, in the initiation phase of mammary carcinogenesis, chemopreventive effects of D-limonene are potentially due to the induction of Phase II carcinogen-metabolizing enzymes, thereby neutralizing the toxicity of chemical carcinogens, and in the postinitiation phase, tumor suppressive activity of D-limonene might be induced by inhibiting the isoprenylation of cell growth-regulating proteins such as Ras and apoptosis induction [79]. The outcome of such inhibitions may alter signal transduction followed by changes in gene expression [98]. For instance, mammary regressing tumors were associated with an increase in the mannose 6-phosphate/IGF II receptor and TGF-β expression in Fischer female rats were placed on a D-limonene diet regimen [98]. These changes can lead to cell cycle arrest in G1 phase, which is followed, by apoptosis, redifferentiation, and finally tumor suppression [99,100]. These findings were recently confirmed by an open label pilot clinical study in patients with breast cancer. In this study, orally administered D-limonene induced downregulation of cyclin D1 expression with subsequent reduced cell proliferation and cell cycle arrest in breast tumor [101]. Different signaling pathways targets of limonene have been reviewed in Fig. 2.4.

On the other hand, it has been demonstrated that oral administration of D-limonene hampers hepato-carcinogenesis induced by *N*-nitrosomorpholine in male Sprague-Dawley rats through induction of apoptosis, inhibition of cell proliferation, and downregulation of oncogene expression [85,102]. This notion is supported by the observation that D-limonene reduces the number of viable LS174T human colon cancer cells *in vitro* by a dose-dependent apoptotic cell death, accounting for its mechanism of action [103]. They observed upregulation of BAX protein, downregulation of Bcl-2 protein, and increase in cytosol cytochrome *c* along with activation of caspase-3, -9 and poly ADP ribose polymerase (PARP) cleavage in a dose-dependent manner in which all are characteristics of apoptosis [103]. Besides, they detected decreased levels of p-Akt (Ser473), p-Akt (Thr308), and p-glycogen synthase kinase-3beta (Ser9) in response to D-limonene treatment [103]. These data may witness that D-limonene induces apoptosis via phosphatidylinositol-4,5-bisphosphate 3-kinase (PI3K)/Akt pathway suppression and mitochondrial death pathway. Another study also has reported the same mechanism for D-limonene on the human leukemia cells (K562 and HL60 cell lines) [104]. Another study revealed highly interesting results as it showed that docetaxel-D-limonene combination can effectively improve the survival of patients with androgen-independent prostate cancer [84]. Results of this study showed that D-limonene enhanced the antitumor effect of docetaxel against prostate cancer in DU-145 cells, generates a greater amount of reactive oxygen species (ROS), and induces apoptotic cell death while having no toxic impact on normal prostate epithelial cells. It also caused a series of changes in different molecules including cytochrome *c* release, cleavages of caspase-9 and -3, and a shift in Bad:Bcl-xL ratio in favor of apoptosis [84]. This combination downregulates several genes involved in cell proliferation (e.g., Bcl-xL), mitotic spindle formation, transcription factors, and oncogenesis whereas upregulates other genes involved in the induction of apoptosis and cell cycle arrest in prostate cancer cells [105]. Taken together, these events significantly enhanced apoptosis and increased cytotoxicity. Consistent with the previous study, another study showed that the combination of D-limonene and linalyl acetate caused caspase-3 activation, PARP cleavage, DNA fragmentation, and cell shrinkage in human SH-SY5Y neuroblastoma cultures, thus suggesting a major role for D-limonene/linalyl acetate combination in cell demise

[106]. Altogether, these results show that mitochondria-mediated intrinsic death pathway might have a prominent role in D-limonene-induced death.

In another study, increased survival of lymphoma-bearing mice fed with a D-limonene diet has demonstrated. They reported increased production of nitric oxide in peritoneal macrophages, delayed phagocytosis and microbicidal activity that imply on immune-modulating properties for D-limonene [107]. Although it has been shown that D-limonene can reduce the tumor size and relapse in a patient with presacral recurrence of an adenocarcinoma in colon but a high number of carcinomas recurred when the D-limonene administration was terminated [28]. This might reflect the temporary effect of D-limonene and the necessity of continual administration to prevent the recurrence of the disease.

Uedo and colleagues orally gave rats chow pellets containing 1% or 2% D-limonene after 25 weeks of the carcinogen treatment and demonstrated that D-limonene attenuated the severity of gastric cancer by enhancing apoptosis, while at the same time decreases the activity of ornithine decarboxylase and DNA synthesis in cancer cells [90]. Metastatic model of human gastric cancer represented a significant reduction in the weight of gastric cancer tumors, downregulation of VEGF, increased apoptotic index and inhibition of liver metastasis in comparison with the control group in response to D-limonene consumption [89]. Recent *in vitro* study verified D-limonene dose-dependent induction of BAX/Bcl, inhibition of VEGF, antimetastasis activity and blockage of VEGF receptor-1 (VEGFR1), activation of p38 and Akt inhibition, decreased expression of matrix metalloproteinases 9 [108]. Vandresen *et al.* synthesized a group of thiosemicarbazones derived from bulky group *R*-(þ-D-limonene) and evaluated *in vitro* effects of these compounds against several human cancer cell lines. They observed excellent growth inhibitory activity against most of the cell lines [109]. This imply on the effectiveness of all isomers of limonene on the treatment of cancer.

Recently, it has been shown that topical treatment of D-limonene inhibits the Ras-ERK signaling pathway, inflammation, and oxidative stress as well as the induction of proapoptotic state in the 7,12-dimethylbenz [α] anthracene-induced skin cancer mouse model [94]. All these changes lead the cell toward apoptosis and preventing the progression of cancer cells [89,104]. A concise yet rather complete description of D-limonene role along with its various mechanisms in different types of cancer is depicted in Table 2.2.

Table 2.2 Summary of Preclinical Studies of Limonene in Cancer Therapy

Type of Cancer	Route of Administration	Type of Study	Mechanism of Action	Reference
Apoptosis pathway as the main mechanism				
Hepatocarcinoma	Oral	Sprague-Dawley rat	Induction of apoptosis; inhibition of cell proliferation; downregulation of oncogene expression	[85]
Colon cancer	Treatment	LS174T cells	Mitochondrial death pathway; suppression of the PI3K/Akt pathway	[103]
Human colon adenocarcinoma	Treatment	HT-29, SW480 and HUVE cells	Induction of apoptosis; inhibition of angiogenesis	[108]
Gastric cancer	Oral	Nude mice	Increased apoptotic index; inhibition of angiogenesis	[89]
		Wistar rat	Induction of apoptosis; decreased DNA synthesis	[90]
Glioma, breast, ovary, kidney, melanoma, NSCLC, prostate colon, leukemia	Treatment	U251, UACC-62, HT-29, MCF-7, PC3, NCI-H460, 746-0, NCI-ADR/RES, OVCAR-03, K562 cells	Cell growth inhibition	[109]
Prostate cancer	Treatment (combined With docetaxel)	DU-145 cells	Induction of apoptosis; Increased ROS generation	[84]

Continued

Table 2.2 Summary of Preclinical Studies of Limonene in Cancer Therapy—cont'd

Type of Cancer	Route of Administration	Type of Study	Mechanism of Action	Reference
Human neuroblastoma	Treatment (combined With linalyl acetate)	SH-SY5Y cells	Apoptotic cell death; necrotic cell death	[106]
Human leukemia cells	Treatment	K562 and HL60 cell	Mitochondrial death pathway	[104]
Inhibition of prenylation as the main mechanism				
Human mammary epithelial	Treatment	NIH3T3 cells	Isoprenylation inhibition of p21ras	[77]
Normal model	Oral	Rat brain	Inhibition of the isoprenylating enzymes	[95]
Skin cancer	Topical	Mouse	Inhibition of farnesylation; activation of ERK pathway; oxidative stress	[94]
Other mechanisms				
Liver cancer	Oral	F344 rat	Inhibition of adducts formation	[80]
Mammary carcinoma	Oral	Fischer rat	Increased expression of the M6P/IGF II receptor and TGF-β	[98]
Mammary carcinomas	Oral	Wistar-Furth rats	Remodeling/ redifferentiation	[91]

3.4. Limonene Biosafety and Adverse Effects

Considering safety matters, Phase I/II clinical trials have been accomplished to ensure the chemical safety of D-limonene oral clinical therapeutic dose [71]. D-limonene has been tested for carcinogenicity in mice and rats and chromosomal aberrations or sister chromatid exchange in cultured

CHO cells and considered to be of fairly low toxicity [110]. Also, no cell mutations were caused by D-limonene in the liver or kidney of rats [111]. Additionally, a clinical study conducted to assess the safety of topically applied D-limonene to the breast of healthy women, found to be safe and feasible for daily application [78]. At pharmacotherapeutic dose suitable for treatment of cancer, limonene showed no toxicity [72]. Despite beneficial effect of limonene some adverse effects have been reported like nephropathy that just occur in male rats, it is proposed that kidney cells of male rats contain high levels of alpha 2u-globulin in kidneys which bind noncovalently to limonene and consequent accumulation of protein in the renal tubules [112]. Another disadvantage of limonene is skin irritation or sensitizing following topical use of limonene although in cosmetic product [15]. Taken together, limonene is a safe drug with no serious risk in therapeutic dose [28,78,15].

4. CONCLUDING REMARKS

POH can be considered as a chemoprotective agent in preclinical studies, and it is currently being evaluated as a clinical candidate to assess the pharmacokinetics and toxicity of POH in human. Because POH interferes with some signaling pathways involved in the pathogenesis of cancers, it makes sense that this natural product can use as a pharmacological-based approach for a potential treatment of human tumors. POH regulates apoptosis pathway, Ras pathway, growth factors, AP-1 activity, differentiation, cell growth, cell cycle, and some other pathways in order to exert its antitumor activity [113,114]. Yet no randomized clinical trials have been conducted to measure the efficacy of POH in human cancer. However, several *in vitro* and *in vivo* investigations have explored this possibility [114]. POH effectiveness may be secondary to any of several factors. Therefore, a further study with a large patient sample is needed for its translation to the clinic.

D-Limonene is an effective, nontoxic dietary antitumor agent that acts through a variety of mechanisms. It holds great promise as a novel class of antitumor drugs for prevention and treatment of different types of cancers [108]. Since in recent years, the effectiveness of D-limonene in cancer prevention and therapy has investigated and understood in different studies, clinical trials started to evaluate its applications. These have shown that it could be considered as a potential chemopreventive candidate/agent. D-limonene has been proved the ability of blocking the growth of many types of cancers through the modulation of diverse molecular target as mentioned above. D-limonene is the first agent that both specifically inhibits protein isoprenylation in cells and has antitumor activity *in vivo* [91], thereby inhibiting tumor growth and metastasis. Also, unlike POH, D-limonene is bioavailable in systemic circulation of human after oral administration [71] and distributes favorably to adipose and mammary tissues in rodents, likely due to its high lipophilicity [78]. Moreover, neglectable toxicity is another important characteristic of D-limonene [91]. D-limonene can be hydroxylated at different positions to augment the chemopreventive activity of this compound, for example, the natural hydroxyl derivatives of this compound (POH) is around 5–10 times more potent than this compound while has the same therapeutic index [99]. More importantly, combination of D-limonene as a novel and potent adjuvant with cytotoxic agents may be more effective as aforementioned in docetaxel example. In the last, it should be emphasized that limonene has transient effect and for effective anticancer strategy continuous administration is mandatory.

Overall, D-limonene has the potential to provide new opportunities for scientists to explore novel therapies against cancer with minimum side effects, so it might be considered for cancer prevention and therapies in future. However, more clinical trials studies warranted establishing D-limonene's role in cancer prevention or treatment.

ACKNOWLEDGMENT

S. G. is supported by University of Manitoba Start up fund.

REFERENCES

[1] V. Tyler, Pharmacognosy, Lea & Febiger, Philadelphia, 1988.
[2] G.J. Kelloff, C.W. Boone, J.A. Crowell, V.E. Steele, R.A. Lubet, L.A. Doody, W.F. Malone, E.T. Hawk, C.C. Sigman, New agents for cancer chemoprevention, J. Cell. Biochem. Suppl. 26 (1996) 1–28.
[3] J.T. Belanger, Perillyl alcohol: Applications in oncology, Altern. Med. Rev. 3 (1998) 448–457.
[4] B. Berman, S. Amini, W. Valins, S. Block, Pharmacotherapy of actinic keratosis, Expert Opin. Pharmacother. 10 (2009) 3015–3031.
[5] R. Prado, S.O. Francis, M.N. Mason, G. Wing, R.G. Gamble, R. Dellavalle, Non-melanoma skin cancer chemoprevention, Dermatol. Surg. 37 (2011) 1566–1578.
[6] P.L. Crowell, M.N. Gould, Chemoprevention and therapy of cancer by d-limonene, Crit. Rev. Oncogen. 5 (1994) 1–22.
[7] H. Loutrari, M. Hatziapostolou, V. Skouridou, E. Papadimitriou, C. Roussos, F.N. Kolisis, A. Papapetropoulos, Perillyl alcohol is an angiogenesis inhibitor, J. Pharmacol. Exper. Therap. 311 (2004) 568–575.
[8] S. Sultana, S. Nafees, A. Khan, Perillyl alcohol as a protective modulator against rat hepatocarcinogenesis via amelioration of oxidative damage and cell proliferation, Hum. Exp. Toxicol. 32 (2013) 1179–1192.

[9] W.C. Margaret Barthelman, Helen L. Gensler, et al., Inhibitory effects of perillyl alcohol on UVB-induced murine skin cancer and AP-1 transactivation, Cancer Res. 58 (1998) 711–716.
[10] D.D. Rovira, The Dictionary of Flavors, Food & Nutrition Press, Inc., Trumbull, USA, 1999.
[11] B. Ruiz, X. Flotats, Citrus essential oils and their influence on the anaerobic digestion process: An overview, Waste Manag. (2014).
[12] W.A. Duetz, H. Bouwmeester, J.B. van Beilen, B. Witholt, Biotransformation of limonene by bacteria, fungi, yeasts, and plants, Appl. Microbiol. Biotechnol. 61 (2003) 269–277.
[13] J.L. Simonsen, The Terpenes, second ed., Vol. 1, Cambridge University Press, 1947, pp. 230–249.
[14] Some chemicals that cause tumours of the kidney or urinary bladder in rodents and some other substances, IARC Monographs on the Evaluation of Carcinogenic Risks to Humans, IARC LYON, France, 73 (1999) 1–674.
[15] Y.W. Kim, M.J. Kim, B.Y. Chung, Y. Bang du, S.K. Lim, S.M. Cho, D.S. Lim, M.C. Cho, K. Yoon, H.S. Kim, K.B. Kim, Y.S. Kim, S.J. Kwack, B.M. Lee, Safety evaluation and risk assessment of d-Limonene, J. Toxicol. Environ. Health B Crit. Rev. 16 (2013) 17–38.
[16] S. Budavari (Ed.), The Merck index- An encyclopedia of chemical, drugs, and biologicals, 12th ed., Merck and Co., Inc., Whitehouse Station, NJ, 1996, p. 413
[17] D.R. Lide, CRC Handbook of Chemistry and Physics, CRC Press, Boca Raton, 72 (1991).
[18] H. Pakdel, Production of dl-limonene by vacuum pyrolysis of used tires, J. Anal. Appl. Pyrol. 57 (2001) 91–107.
[19] A.T. Karlberg, K. Magnusson, U. Nilsson, Air oxidation of d-limonene (the citrus solvent) creates potent allergens, Contact Dermatitis 26 (1992) 332–340.
[20] R.K. Pathak, K. Salo, E.U. Emanuelsson, C. Cai, A. Lutz, A.M. Hallquist, M. Hallquist, Influence of ozone and radical chemistry on limonene organic aerosol production and thermal characteristics, Environ. Sci. Technol. 46 (2012) 11660–11669.
[21] R.S. Dhavalikar, P.K. Bhattacharyya, Microbiological trans- formations of terpenes. VIII. Fermentation of limonene in a soil pseudomonad, Indian J. Biochem. 3 (1966) 144–157.
[22] O.P. Shulka, P.K. Bhattacharyya, Microbial transformation of terpenes: Part XI—Pathways of degradation of alpha & beta pinenes in a soil pseudomonad (PL-strain), Indian J. Biochem. 5 (1968) 92–101.
[23] Chemical Information Services, Directory of World Chemical Producers 1995/96 Standard Edition, Chemical Information Services Inc., Dallas, TX, 1995, p.452.
[24] A.Z.M. Badee, S.A. Helmy, N.F.S. Morsy, Utilisation of orange peel in the production of -terpineol by *Penicillium digitatum* (NRRL 1202), Food Chem. 126 (2011) 849–854.
[25] F. Bakkali, S. Averbeck, D. Averbeck, M. Idaomar, Biological effects of essential oils—A review, Food Chem. Toxicol. 46 (2008) 446–475.
[26] C.C.C.R. de Carvalho, M.M.R. da Fonseca, Biotransformation of terpenes, Biotechnol. Adv. 24 (2006) 134–142.
[27] M.H. Pan, C.T. Ho, Chemopreventive effects of natural dietary compounds on cancer development, Chem. Soc. Rev. 37 (2008) 2558–2574.
[28] J. Sun, D-Limonene: Safety and clinical applications, Altern. Med. Rev. 12 (2007) 259–264.
[29] M. Miyazawa, M. Shindo, T. Shimada, Metabolism of (+)- and (-)-limonenes to respective carveols and perillyl alcohols by CYP2C9 and CYP2C19 in human liver microsomes, Drug Metab. Dispos. 30 (2002) 602–607.

[30] J.L. Bicas, P. Fontanille, G.M. Pastore, C. Larroche, Characterization of monoterpene biotransformation in two pseudomonads, J. Appl. Microbiol. 105 (2008) 1991–2001.
[31] M. van der Werf, J. de Bont, D. Leak, Opportunities in microbial biotransformation of monoterpenes, Adv. Biochem. Eng. Biotechnol. 55 (1997) 147–177.
[32] E.A. Ariazi, Y. Satomi, M.J. Ellis, J.D. Haag, W. Shi, C.A. Sattler, M.N. Gould, Activation of the transforming growth factor beta signaling pathway and induction of cytostasis and apoptosis in mammary carcinomas treated with the anticancer agent perillyl alcohol, Cancer Res. 59 (1999) 1917–1928.
[33] M. Barthelman, W. Chen, H.L. Gensler, C. Huang, Z. Dong, G.T. Bowden, Inhibitory effects of perillyl alcohol on UVB-induced murine skin cancer and AP-1 transactivation, Cancer Res. 58 (1998) 711–716.
[34] P.L. Crowell, Z. Ren, S. Lin, E. Vedejs, M.N. Gould, Structure-activity relationships among monoterpene inhibitors of protein isoprenylation and cell proliferation, Biochem. Pharmacol. 47 (1994) 1405–1415.
[35] J.J. Mills, R.S. Chari, I.J. Boyer, M.N. Gould, R.L. Jirtle, Induction of apoptosis in liver tumors by the monoterpene perillyl alcohol, Cancer Res. 55 (1995) 979–983.
[36] W. Shi, M.N. Gould, Induction of differentiation in neuro-2A cells by the monoterpene perillyl alcohol, Cancer Lett. 95 (1995) 1–6.
[37] W. Shi, M.N. Gould, Induction of cytostasis in mammary carcinoma cells treated with the anticancer agent perillyl alcohol, Carcinogenesis 23 (2002) 131–142.
[38] H. Mo, C.E. Elson, Studies of the isoprenoid-mediated inhibition of mevalonate synthesis applied to cancer chemotherapy and chemoprevention, Exp. Biol. Med. (Maywood) 229 (2004) 567–585.
[39] M.M. Feldkamp, N. Lau, A. Guha, Growth inhibition of astrocytoma cells by farnesyl transferase inhibitors is mediated by a combination of anti-proliferative, pro-apoptotic and anti-angiogenic effects, Oncogene 18 (1999) 7514–7526.
[40] C.O. da Fonseca, J.A. Landeiro, S.S. Clark, T. Quirico-Santos, G. da Costa Carvalho Mda, C.R. Gattass, Recent advances in the molecular genetics of malignant gliomas disclose targets for antitumor agent perillyl alcohol, Surg. Neurol. 65 (Suppl. 1) (2006) S1:2–S1:8, discussion S1:8-1:9.
[41] K.R. Stayrook, J.H. McKinzie, Y.D. Burke, Y.A. Burke, P.L. Crowell, Induction of the apoptosis-promoting protein Bak by perillyl alcohol in pancreatic ductal adenocarcinoma relative to untransformed ductal epithelial cells, Carcinogenesis 18 (1997) 1655–1658.
[42] S.S. Clark, L. Zhong, D. Filiault, S. Perman, Z. Ren, M. Gould, X. Yang, Antileukemia effect of perillyl alcohol in Bcr/Abl-transformed cells indirectly inhibits signaling through Mek in a Ras-and Raf-independent fashion, Clin. Cancer Res. 9 (2003) 4494–4504.
[43] H. Loutrari, M. Hatziapostolou, V. Skouridou, E. Papadimitriou, C. Roussos, F.N. Kolisis, A. Papapetropoulos, Perillyl alcohol is an angiogenesis inhibitor, J. Pharmacol. Exp. Ther. 311 (2004) 568–575.
[44] D. Rajesh, R.A. Stenzel, S.P. Howard, Perillyl alcohol as a radio-/chemosensitizer in malignant glioma, J. Biol. Chem. 278 (2003) 35968–35978.
[45] Y. Satomi, S. Miyamoto, M.N. Gould, Induction of AP-1 activity by perillyl alcohol in breast cancer cells, Carcinogenesis 20 (1999) 1957–1961.
[46] F. Colotta, N. Polentarutti, M. Sironi, A. Mantovani, Expression and involvement of c-fos and c-jun protooncogenes in programmed cell death induced by growth factor deprivation in lymphoid cell lines, J. Biol. Chem. 267 (1992) 18278–18283.
[47] H. Sawai, T. Okazaki, H. Yamamoto, H. Okano, Y. Takeda, M. Tashima, H. Sawada, M. Okuma, H. Ishikura, H. Umehara, et al., Requirement of AP-1 for ceramide-induced apoptosis in human leukemia HL-60 cells, J. Biol. Chem. 270 (1995) 27326–27331.

[48] M. Watabe, K. Ito, Y. Masuda, S. Nakajo, K. Nakaya, Activation of AP-1 is required for bufalin-induced apoptosis in human leukemia U937 cells, Oncogene 16 (1998) 779–787.
[49] C. Gajate, A. Santos-Beneit, M. Modolell, F. Mollinedo, Involvement of c-Jun NH2-terminal kinase activation and c-Jun in the induction of apoptosis by the ether phospholipid 1-O-octadecyl-2-O-methyl-rac-glycero-3-phosphocholine, Mol. Pharmacol. 53 (1998) 602–612.
[50] B.H. Chung, H.Y. Lee, J.S. Lee, C.Y. Young, Perillyl alcohol inhibits the expression and function of the androgen receptor in human prostate cancer cells, Cancer Lett. 236 (2006) 222–228.
[51] S. Bardon, K. Picard, P. Martel, Monoterpenes inhibit cell growth, cell cycle progression, and cyclin D1 gene expression in human breast cancer cell lines, Nutr. Cancer 32 (1998) 1–7.
[52] J.A. Elegbede, R. Flores, R.C. Wang, Perillyl alcohol and perillaldehyde induced cell cycle arrest and cell death in BroTo and A549 cells cultured in vitro, Life Sci. 73 (2003) 2831–2840.
[53] T. Yuri, N. Danbara, M. Tsujita-Kyutoku, Y. Kiyozuka, H. Senzaki, N. Shikata, H. Kanzaki, A. Tsubura, Perillyl alcohol inhibits human breast cancer cell growth in vitro and in vivo, Breast Cancer Res. Treat. 84 (2004) 251–260.
[54] T. Balassiano, A. de Paulo, H. Silva, C. Cabral, O. da Fonseca, D.C.C. Da Glória, Effects of perillyl alcohol in glial C6 cell line in vitro and anti-metastatic activity in chorioallantoic membrane model, Int. J. Mol. Med. 10 (2002) 785–788.
[55] J.E. Wagner, J.L. Huff, W.L. Rust, K. Kingsley, G.E. Plopper, Perillyl alcohol inhibits breast cell migration without affecting cell adhesion, BioMed Res. Inter. 2 (2002) 136–140.
[56] T. Sundin, D.M. Peffley, P. Hentosh, Disruption of an hTERT-mTOR-RAPTOR protein complex by a phytochemical perillyl alcohol and rapamycin, Mol. Cell. Biochem. 375 (2013) 97–104.
[57] T. Sundin, D. Peffley, P. Hentosh, eIF4E-Overexpression imparts perillyl alcohol and rapamycin-mediated regulation of telomerase reverse transcriptase, Exp. Cell Res. 319 (2013) 2103–2112.
[58] L.R. Phillips, L. Malspeis, J.G. Supko, Pharmacokinetics of active drug metabolites after oral administration of perillyl alcohol, an investigational antineoplastic agent, to the dog, Drug Metab. Dispos. 23 (1995) 676–680.
[59] E. Evans, D. Arneson, R. Kovatch, J. Supko, T. Morton, L. Siemann, J. Cannon, J. Tomaszewski, A. Smith, Toxicology and pharmacology of perillyl alcohol (NSC-641066) in rats and dogs, Proc. Am. Assoc. Cancer Res. 36 (1995) 366.
[60] G.H. Ripple, M.N. Gould, J.A. Stewart, K.D. Tutsch, R.Z. Arzoomanian, D. Alberti, C. Feierabend, M. Pomplun, G. Wilding, H.H. Bailey, Phase I clinical trial of perillyl alcohol administered daily, Clin. Cancer Res. 4 (1998) 1159–1164.
[61] G.H. Ripple, M.N. Gould, R.Z. Arzoomanian, D. Alberti, C. Feierabend, K. Simon, K. Binger, K.D. Tutsch, M. Pomplun, A. Wahamaki, R. Marnocha, G. Wilding, H.H. Bailey, Phase I clinical and pharmacokinetic study of perillyl alcohol administered four times a day, Clin. Cancer Res. 6 (2000) 390–396.
[62] G.R. Hudes, C.E. Szarka, A. Adams, S. Ranganathan, R.A. McCauley, L.M. Weiner, C.J. Langer, S. Litwin, G. Yeslow, T. Halberr, M. Qian, J.M. Gallo, Phase I pharmacokinetic trial of perillyl alcohol (NSC 641066) in patients with refractory solid malignancies, Clin. Cancer Res. 6 (2000) 3071–3080.
[63] C.G. Azzoli, V.A. Miller, K.K. Ng, L.M. Krug, D.R. Spriggs, W.P. Tong, E.R. Riedel, M.G. Kris, A phase I trial of perillyl alcohol in patients with advanced solid tumors, Cancer Chemother. Pharmacol. 51 (2003) 493–498.

[64] H.H. Bailey, D. Levy, L.S. Harris, J.C. Schink, F. Foss, P. Beatty, S. Wadler, A phase II trial of daily perillyl alcohol in patients with advanced ovarian cancer: Eastern Cooperative Oncology Group Study E2E96, Gynecol. Oncol. 85 (2002) 464–468.
[65] G. Liu, K. Oettel, H. Bailey, L.V. Ummersen, K. Tutsch, M.J. Staab, D. Horvath, D. Alberti, R. Arzoomanian, H. Rezazadeh, J. McGovern, E. Robinson, D. DeMets, G. Wilding, Phase II trial of perillyl alcohol (NSC 641066) administered daily in patients with metastatic androgen independent prostate cancer, Invest. New Drugs 21 (2003) 367–372.
[66] S.M. Meadows, D. Mulkerin, J. Berlin, H. Bailey, J. Kolesar, D. Warren, J.P. Thomas, Phase II trial of perillyl alcohol in patients with metastatic colorectal cancer, Int. J. Gastrointest. Cancer 32 (2002) 125–128.
[67] H.H. Bailey, S. Attia, R.R. Love, T. Fass, R. Chappell, K. Tutsch, L. Harris, A. Jumonville, R. Hansen, G.R. Shapiro, J.A. Stewart, Phase II trial of daily oral perillyl alcohol (NSC 641066) in treatment-refractory metastatic breast cancer, Cancer Chemother. Pharmacol. 62 (2008) 149–157.
[68] H. Genchi, K. Mori, Evaluation and feature extraction on automatic pattern recognition system, Denki Tsuchin Gakkai Pari (in Japanese) 1 (1965) 1–6.
[69] S.P. Stratton, K.L. Saboda, P.B. Myrdal, A. Gupta, N.E. McKenzie, C. Brooks, S.J. Salasche, J.A. Warneke, J. Ranger-Moore, P.D. Bozzo, J. Blanchard, J.G. Einspahr, R.T. Dorr, N. Levine, D.S. Alberts, Phase 1 study of topical perillyl alcohol cream for chemoprevention of skin cancer, Nutr. Cancer 60 (2008) 325–330.
[70] C.O. da Fonseca, G. Schwartsmann, J. Fischer, J. Nagel, D. Futuro, T. Quirico-Santos, C.R. Gattass, Preliminary results from a phase I/II study of perillyl alcohol intranasal administration in adults with recurrent malignant gliomas, Surg. Neurol. 70 (2008) 259–266, discussion 266–257.
[71] D.M. Vigushin, G.K. Poon, A. Boddy, J. English, G.W. Halbert, C. Pagonis, M. Jarman, R.C. Coombes, Phase I and pharmacokinetic study of D-limonene in patients with advanced cancer. Cancer Research Campaign Phase I/II Clinical Trials Committee, Cancer Chemother. Pharmacol. 42 (1998) 111–117.
[72] P.L. Crowell, C.E. Elson, H.H. Bailey, A. Elegbede, J.D. Haag, M.N. Gould, Human metabolism of the experimental cancer therapeutic agent d-limonene, Cancer Chemother. Pharmacol. 35 (1994) 31–37.
[73] G.K. Poon, D. Vigushin, L.J. Griggs, M.G. Rowlands, R.C. Coombes, M. Jarman, Identification and characterization of limonene metabolites in patients with advanced cancer by liquid chromatography/mass spectrometry, Drug Metab. Dispos. 24 (1996) 565–571.
[74] P.L. Crowell, S. Lin, E. Vedejs, M.N. Gould, Identification of metabolites of the antitumor agent d-limonene capable of inhibiting protein isoprenylation and cell growth, Cancer Chemother. Pharmacol. 31 (1992) 205–212.
[75] G.D. Bell, Cholelitholytic agents, Pharmacol. Ther. 23 (1983) 79–108.
[76] P.L. Crowell, M.N. Gould, Chemoprevention and therapy of cancer by d-limonene, Crit. Rev. Oncog. 5 (1994) 1–22.
[77] P.L. Crowell, R.R. Chang, Z.B. Ren, C.E. Elson, M.N. Gould, Selective inhibition of isoprenylation of 21-26-kDa proteins by the anticarcinogen d-limonene and its metabolites, J. Biol. Chem. 266 (1991) 17679–17685.
[78] J.A. Miller, P.A. Thompson, I.A. Hakim, A.M. Lopez, C.A. Thomson, W. Chew, C.H. Hsu, H.H. Chow, Safety and feasibility of topical application of limonene as a massage oil to the breast, J. Cancer Ther. 3 (2012).
[79] P.L. Crowell, Prevention and therapy of cancer by dietary monoterpenes, J. Nutr. 129 (1999) 775S–778S.
[80] J.A. Elegbede, M.N. Gould, Monoterpenes reduced adducts formation in rats exposed to aflatoxin B1, Afr. J. Biotechnol. 1 (2002) 46–49.

[81] D. Roberto, P. Micucci, T. Sebastian, F. Graciela, C. Anesini, Antioxidant activity of limonene on normal murine lymphocytes: Relation to H2O2 modulation and cell proliferation, Basic Clin. Pharmacol. Toxicol. 106 (2010) 38–44.
[82] M. Hotta, R. Nakata, M. Katsukawa, K. Hori, S. Takahashi, H. Inoue, Carvacrol, a component of thyme oil, activates PPARalpha and gamma and suppresses COX-2 expression, J. Lipid Res. 51 (2010) 132–139.
[83] W.J. Yoon, N.H. Lee, C.G. Hyun, Limonene suppresses lipopolysaccharide-induced production of nitric oxide, prostaglandin E2, and pro-inflammatory cytokines in RAW 264.7 macrophages, J. Oleo Sci. 59 (2010) 415–421.
[84] T. Rabi, A. Bishayee, d-Limonene sensitizes docetaxel-induced cytotoxicity in human prostate cancer cells: Generation of reactive oxygen species and induction of apoptosis, J. Carcinog. 8 (2009) 9.
[85] I. Kaji, M. Tatsuta, H. Iishi, M. Baba, A. Inoue, H. Kasugai, Inhibition by d-limonene of experimental hepatocarcinogenesis in Sprague-Dawley rats does not involve p21(ras) plasma membrane association, Int. J. Cancer 93 (2001) 441–444.
[86] L.W. Wattenberg, J.B. Coccia, Inhibition of 4-(methylnitrosamino)-1-(3-pyridyl)-1-butanone carcinogenesis in mice by D-limonene and citrus fruit oils, Carcinogenesis 12 (1991) 115–117.
[87] C. Cardenas, A.R. Quesada, M.A. Medina, Effects of ursolic acid on different steps of the angiogenic process, Biochem. Biophys. Res. Commun. 320 (2004) 402–408.
[88] K.Z. Guyton, T.W. Kensler, Prevention of liver cancer, Curr. Oncol. Rep. 4 (2002) 464–470.
[89] X.G. Lu, L.B. Zhan, B.A. Feng, M.Y. Qu, L.H. Yu, J.H. Xie, Inhibition of growth and metastasis of human gastric cancer implanted in nude mice by d-limonene, World J. Gastroenterol. 10 (2004) 2140–2144.
[90] N. Uedo, M. Tatsuta, H. Iishi, M. Baba, N. Sakai, H. Yano, T. Otani, Inhibition by D-limonene of gastric carcinogenesis induced by N-methyl-N'-nitro-N-nitrosoguanidine in Wistar rats, Cancer Lett. 137 (1999) 131–136.
[91] J.D. Haag, M.J. Lindstrom, M.N. Gould, Limonene-induced regression of mammary carcinomas, Cancer Res. 52 (1992) 4021–4026.
[92] T.H. Maltzman, L.M. Hurt, C.E. Elson, M.A. Tanner, M.N. Gould, The prevention of nitrosomethylurea-induced mammary tumors by d-limonene and orange oil, Carcinogenesis 10 (1989) 781–783.
[93] I.A. Hakim, R.B. Harris, C. Ritenbaugh, Citrus peel use is associated with reduced risk of squamous cell carcinoma of the skin, Nutr. Cancer 37 (2000) 161–168.
[94] M.G. Manuele, M.L. Barreiro Arcos, R. Davicino, G. Ferraro, G. Cremaschi, C. Anesini, Limonene exerts antiproliferative effects and increases nitric oxide levels on a lymphoma cell line by dual mechanism of the ERK pathway: Relationship with oxidative stress, Cancer Invest. 28 (2010) 135–145.
[95] I.R. Hardcastle, M.G. Rowlands, A.M. Barber, R.M. Grimshaw, M.K. Mohan, B.P. Nutley, M. Jarman, Inhibition of protein prenylation by metabolites of limonene, Biochem. Pharmacol. 57 (1999) 801–809.
[96] T.P. Ong, M.T. Cardozo, A. de Conti, F.S. Moreno, Chemoprevention of hepatocarcinogenesis with dietary isoprenic derivatives: Cellular and molecular aspects, Curr. Cancer Drug Targets 12 (2012) 1173–1190.
[97] B. Yeganeh, E. Wiechec, S.R. Ande, P. Sharma, A.R. Moghadam, M. Post, D.H. Freed, M. Hashemi, S. Shojaei, A.A. Zeki, S. Ghavami, Targeting the mevalonate cascade as a new therapeutic approach in heart disease, cancer and pulmonary disease, Pharmacol. Ther. 143 (2014) 87–110.
[98] R.L. Jirtle, J.D. Haag, E.A. Ariazi, M.N. Gould, Increased mannose 6-phosphate/insulin-like growth factor II receptor and transforming growth factor beta 1 levels during monoterpene-induced regression of mammary tumors, Cancer Res. 53 (1993) 3849–3852.

[99] M.N. Gould, Prevention and therapy of mammary cancer by monoterpenes, J. Cell. Biochem. (Suppl. 22) (1995) 139–144.
[100] M.N. Gould, Cancer chemoprevention and therapy by monoterpenes, Environ. Health Perspect. 105 (Suppl. 4) (1997) 977–979.
[101] J.A. Miller, J.E. Lang, M. Ley, R. Nagle, C.H. Hsu, P.A. Thompson, C. Cordova, A. Waer, H.H. Chow, Human breast tissue disposition and bioactivity of limonene in women with early-stage breast cancer, Cancer Prev. Res. (Phila.) 6 (2013) 577–584.
[102] R.K. Giri, T. Parija, B.R. Das, d-Limonene chemoprevention of hepatocarcinogenesis in AKR mice: Inhibition of c-jun and c-myc, Oncol. Rep. 6 (1999) 1123–1127.
[103] S.S. Jia, G.P. Xi, M. Zhang, Y.B. Chen, B. Lei, X.S. Dong, Y.M. Yang, Induction of apoptosis by D-limonene is mediated by inactivation of Akt in LS174T human colon cancer cells, Oncol. Rep. 29 (2013) 349–354.
[104] J. Ji, L. Zhang, Y.Y. Wu, X.Y. Zhu, S.Q. Lv, X.Z. Sun, Induction of apoptosis by d-limonene is mediated by a caspase-dependent mitochondrial death pathway in human leukemia cells, Leuk. Lymphoma 47 (2006) 2617–2624.
[105] F. Fabbri, D. Amadori, S. Carloni, G. Brigliadori, A. Tesei, P. Ulivi, M. Rosetti, I. Vannini, C. Arienti, W. Zoli, R. Silvestrini, Mitotic catastrophe and apoptosis induced by docetaxel in hormone-refractory prostate cancer cells, J. Cell. Physiol. 217 (2008) 494–501.
[106] R. Russo, A. Ciociaro, L. Berliocchi, M.G. Cassiano, L. Rombola, S. Ragusa, G. Bagetta, F. Blandini, M.T. Corasaniti, Implication of limonene and linalyl acetate in cytotoxicity induced by bergamot essential oil in human neuroblastoma cells, Fitoterapia 89 (2013) 48–57.
[107] S. Del Toro-Arreola, E. Flores-Torales, C. Torres-Lozano, A. Del Toro-Arreola, K. Tostado-Pelayo, M. Guadalupe Ramirez-Duenas, A. Daneri-Navarro, Effect of D-limonene on immune response in BALB/c mice with lymphoma, Int. Immunopharmacol. 5 (2005) 829–838.
[108] K.N. Chidambara Murthy, G.K. Jayaprakasha, B.S. Patil, D-limonene rich volatile oil from blood oranges inhibits angiogenesis, metastasis and cell death in human colon cancer cells, Life Sci. 91 (2012) 429–439.
[109] F. Vandresen, H. Falzirolli, S.A. Almeida Batista, A.P. da Silva-Giardini, D.N. de Oliveira, R.R. Catharino, A.L. Ruiz, J.E. de Carvalho, M.A. Foglio, C.C. da Silva, Novel R-(+)-limonene-based thiosemicarbazones and their antitumor activity against human tumor cell lines, Eur. J. Med. Chem. 79 (2014) 110–116.
[110] NTP, NTP toxicology and carcinogenesis studies of d-limonene (CAS No. 5989-27-5) in F344/N rats and B6C3F1 mice (gavage studies), Natl. Toxicol. Program Tech. Rep. Ser. 347 (1990) 1–165.
[111] S.D. Turner, H. Tinwell, W. Piegorsch, P. Schmezer, J. Ashby, The male rat carcinogens limonene and sodium saccharin are not mutagenic to male Big Blue rats, Mutagenesis 16 (2001) 329–332.
[112] M. Miyazawa, M. Shindo, T. Shimada, Sex differences in the metabolism of (+)- and (-)-limonene enantiomers to carveol and perillyl alcohol derivatives by cytochrome p450 enzymes in rat liver microsomes, Chem. Res. Toxicol. 15 (2002) 15–20.
[113] S. Sarkar, B. Azab, B. Quinn, X. Shen, P. Dent, A. Klibanov, L. Emdad, S. Das, D. Sarkar, P. Fisher, Chemoprevention gene therapy (CGT) of pancreatic cancer using perillyl alcohol and a novel chimeric serotype cancer terminator virus, Curr. Mol. Med. 14 (2014) 125–140.
[114] M.M. Silva, C.O. Da Fonseca, R. Moura-Neto, J.F. Carvalho, T. Quirico-Santos, M.G. Carvalho, Influence of GSTM1 and GSTT1 polymorphisms on the survival rate of patients with malignant glioma under perillyl alcohol-based therapy, Genet. Mol. Res. 12 (2013) 1621–1630.

CHAPTER THREE

Ganoderic Acid and Lucidenic Acid (Triterpenoid)

Chin-Lin Hsu*,†, Gow-Chin Yen‡,§,1
*School of Nutrition, Chung Shan Medical University, Taichung, Taiwan
†Department of Nutrition, Chung Shan Medical University Hospital, Taichung, Taiwan
‡Department of Food Science and Biotechnology, National Chung Hsing University, Taichung, Taiwan
§Agricultural Biotechnology Center, National Chung Hsing University, Taichung, Taiwan
[1]Corresponding author: e-mail address: gcyen@dragon.nchu.edu.tw

Contents

Abstract

Ganoderma species is known as a functional mushroom used in many Asian countries. Triterpenoids, ganoderic acids, and lucidenic acids have been isolated from *Ganoderma* species (Ganodermataceae family), such as *Ganoderma lucidum*. Triterpenoids have been investigated for their biological activities, including antibacterial, antiviral, antitumor, antiosteoclastic differentiation activity, anti-HIV-1, hepatoprotection, antioxidation, antihypertension, cholesterol reduction, and antiaggregation functions. In this chapter, the sources, biosynthesis, biological functions (including cell cytotoxicity, cell apoptosis, cell cycle arrest, anti-invasion, autophagy, anti-inflammation, antiosteoclastogenesis, antiasthma, and antihepatitis B activity), and pharmacokinetics of lucidenic acids and ganoderic acids are considered in detail here. This chapter briefly

The Enzymes, Volume 36
ISSN 1874-6047
http://dx.doi.org/10.1016/B978-0-12-802215-3.00003-3

summarizes the multiple functions of lucidenic acids and ganoderic acids and their potential for fighting against human diseases.

1. INTRODUCTION

Triterpenoid-rich fractions were obtained from *Ganoderma* mushroom (Ganodermataceae). Triterpenoids (such as lucidenic acids and ganoderic acids) have been reported to exhibit various biological activities, such as antibacterial, antiviral, antitumor, antiosteoclastic differentiation activity, anti-HIV-1, hepatoprotection, antioxidation, antihypertension, cholesterol reduction, and antiaggregation functions [1–13]. Many reports have indicated that lucidenic acids and ganoderic acids were isolated from a key strain of *Ganoderma lucidum* [10,14–24]. *G. lucidum* is a medicinal mushroom used for several thousand years in Asian countries, such as China (called as "Lingzhi"), Japan (called as "reishi" or "Mannentake"), Taiwan (called as "Lingzhi"), and Korea (called as "Yeongji"). It has been used as a folk remedy for prevention and treatment of various human diseases, including against cancer, immunological disorders, and potential in sustainably enhancing human vitality and promoting longevity.

Lucidenic acids and ganoderic acids are isolated from different sources (including fungus and fruiting body) of Ganodermataceae family, such as *G. lucidum*, *Ganoderma colossum*, *Ganoderma applanatum*, *Ganoderma resinaceum*, and *Ganoderma orbiforme*. These triterpenoids have pharmacological properties, including cell cytotoxicity, cell apoptosis, cell cycle arrest, anti-invasion, autophagy, anti-inflammation, antiosteoclastogenesis, antiasthma, and antihepatitis B activity [10,25–32]. Other biologically active compounds (such as sterols, polysaccharides, and glycoproteins) have also been isolated from these mushrooms and possessed pharmacological activities including cytotoxic, antiviral, anti-inflammatory, hypoglycemic, and immunostimulant activities [33–37]. Many reports have also indicated that the functional compounds from *G. lucidum* showed inhibitions of cell proliferation, angiogenesis, invasion, and metastasis and induced cell cycle arrest and apoptosis in carcinoma cell lines and animal models [38–42]. Moreover, the content of triterpenoids isolated from fruit body of *Ganoderma* species is too low and very expensive. Optimization of the fermentation process (including *Ganoderma* strain screening and optimization of the fermentation process conditions) plays an important role in improving the production of triterpenoids [43–51]. Finally, recent reports have indicated that the pharmacokinetics of ganoderic acids from *Ganoderma* genus *in vitro* and *in vivo* were investigated [52–55]. Triterpenoids' metabolism plays an important role in medicinal plant drug discovery, development, and clinical uses. This chapter presents key presentations from the sources, biosynthesis, biological functions, and pharmacokinetics of triterpenoids (such as lucidenic acids and ganoderic acids) from Ganodermataceae family.

2. LUCIDENIC ACIDS AND GANODERIC ACIDS FROM *GANODERMA* SPECIES

2.1. The Sources of Lucidenic Acids and Ganoderic Acids

Many triterpenoids (including lucidenic acids and ganoderic acids) were isolated from different parts (including fungus and fruiting body) of *Ganoderma* species, such as *G. lucidum*, *G. colossum*, *G. applanatum*, *G. resinaceum*, and *G. orbiforme*. In *G. lucidum* strain, Mizushina *et al.* [14] indicated that lucidenic acid O, lucidenic lactone, and cerevisterol were isolated from the fruiting caps of the mushroom *G. lucidum*. Gao *et al.* [15] indicated that lanostante-type triterpene aldehydes, such as lucialdehydes A, B, and C, ganodermanonol, ganodermadiol, ganodermanondiol, ganodermanontriol (GDNT), and ganoderic acids A, B8, and C1, were isolated from the fruiting bodies of *G. lucidum*. Iwatsuki *et al.* [16] indicated that lucidenic acids (including A, C, D2, E2, and F), ganoderic acids (including E, F, and T–Q), methyl lucidenates (including A, D2, E2, F, and L), and methyl ganoderate F were isolated and identified from the fungus *G. lucidum*. The ethanol extract from the fruiting bodies of *G. lucidum* Karst. led to the isolation of triterpenoids, which were ganoderic acid DM, methyl ester, ganoderic acid D, ganoderic acid A, ganodermadiol, lucidumol B, GDNT, and ganoderiol A [17]. Weng *et al.* [56] and Hsu *et al.* [10] indicated that the lucidenic acids A, B, C, and N were isolated from a new *G. lucidum* strain (YK-02). Akihisa *et al.* [19] indicated that the fungus *G. lucidum* led to the isolation of lucidenic acids (including lucidenic acid B, methyl lucidenate C, lucidenic acid F, lucidenic acid N, 20(21)-dehydrolucidenic acid A, methyl 20(21)-dehydrolucidenate A, 20-hydroxylucidenic acid D2, 20-hydroxylucidenic acid E2, 20-hydroxylucidenic acid F, 20-hydroxylucidenic acid N, and 20-hydroxylucidenic acid P), ganoderic acids (including ganoderic acid A, ganoderic acid C1, ganoderic acid C2, ganoderic acid DM, ganodermanondiol, and ganolactone), and sterols (including fungisterol, 5,6-dihydroergosterol, ergosterol, ergosterol peroxide, 9(11)-dehydroergosterol peroxide, and demethylincisterol A3). Seo *et al.* [20] indicated that steroids (including ergosterol peroxide, ergosterol, and stella sterol) and triterpenes (including ganoderic acid Sz, ganoderic acid C1, ganoderic acid A, methyl ganoderate A, and lucidenic acid A) were isolated from the fruiting bodies of *G. lucidum*. Lee *et al.* [21] indicated that lanostane triterpenes (including methyl ganoderate A acetonide, *n*-Butyl ganoderate H, methyl ganoderate A, ganoderic acid B, ganoderic acid E, ganolucidic acid A, ganodermadiol, ganoderic acid Y, ganoderiol F, lucidumol B, ganodermanondiol, GDNT, lucidadiol, lucidenic acid N, lucidenic acid A, methyl lucidenate E2, *n*-Butyl lucidenate N, and *n*-Butyl lucidenate A) were isolated from fruiting bodies of *G. lucidum*. Grienke *et al.* [22] indicated that bioactive triterpenes (including ganoderol B, ganodermanondiol, ganoderic acid TR, GDNT, ganoderiol F, ganoderic acid DM, ganoderon B, lucidumol A, lucidenic acid A, ganoderic acid D, ganoderic acid F, ganoderenic acid D, ganoderic acid J, ganoderic acid B, ganoderic acid A, methyl ganoderate H, ganoderic acid H, and ganoderic acid AM1) were isolated from the dried fruiting bodies of *G. lucidum*. Ding *et al.* [23] indicated that ganoderic acids T, Mk, Me, and S were isolated from dried fermentation mycelia powder of *G. lucidum*. Li *et al.* [24] also indicated that ganoderic acids (including 3α,22β-diacetoxy-7α-hydroxyl-5α-lanost-8,24*E*-dien-26-oic acid, ganoderic acids Mk, Mc, S, and Mf) were isolated from *G. lucidum* mycelia.

In the other *Ganoderma* species strains, the ethanol extract from the fruiting bodies of *G. colossum* contains lucidenic acids A and B [28]. Some triterpenoids, including ganoderic acids AP2 and AP3 and ganoderenic acids A, B, D, and G, were isolated from the fungus *G. applanatum* [57]. Peng *et al.* [58] indicated that lanostanoids, named lucidones A, B, D, E, F, G, and H, 7-oxo-ganoderic acid Z2, 7-oxo-ganoderic acid Z3, ganoderesin A, ganoderesin B, ganoderic acid Y, ganoderol B, and 7-oxo-ganoderic acid Z were isolated from the fruiting bodies of *G. resinaceum*. Isaka *et al.* [59] indicated that lanostane triterpenes (including ganorbiformins A–G and ganoderic acids V, T, R, P, X, and S) were isolated from the basidiomycete *G. orbiforme* BCC 22324. The chemical structures of some lucidenic acids (LAs) and ganoderic acids (GAs) are shown in Figs. 3.1 and 3.2.

2.2. The Biosynthesis of Ganoderic Acids

The biosynthesis of ganoderic acids plays an important role in the enhancement of bioactive compounds' production. Ganoderic acids found in *G. lucidum* have multiple pharmacological activities, such as anticancer, antiviral, antioxidant, antiplatelet aggregation, inhibitions of histamine release

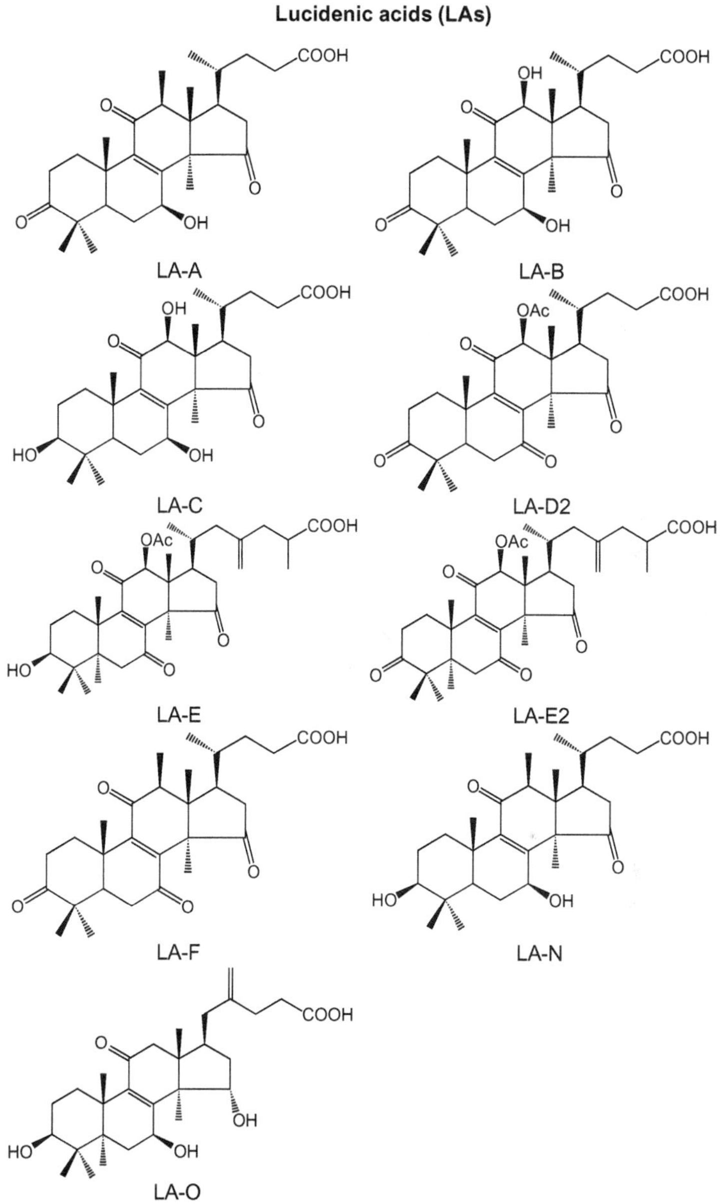

Figure 3.1 Chemical structures of lucidenic acids (LAs).

and cholesterol synthesis, and anti-HIV-1 [60–62]. The biosynthesis of ganoderic acids has been proposed to proceed via the mevalonate/isoprenoid pathway. Shi *et al.* [63] and Ren *et al.* [64] indicated that the modulations of 3-hydroxy-3-methylglutaryl coenzyme A reductase (*HMGR*), squalene

Ganoderic acids (GAs)

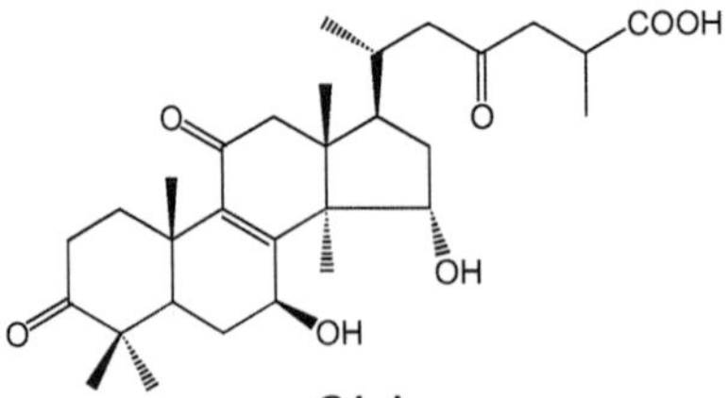

GA-A

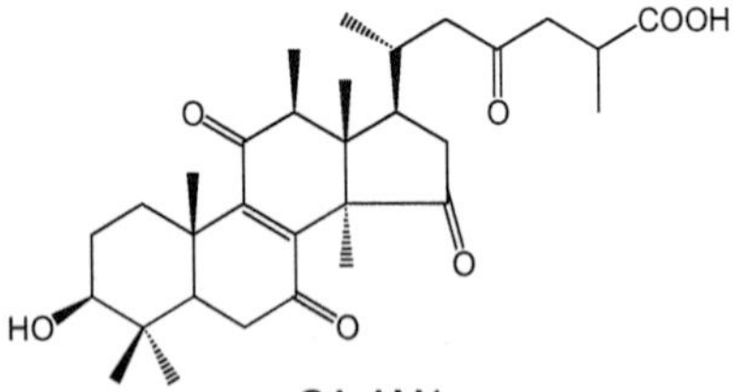

GA-AM1

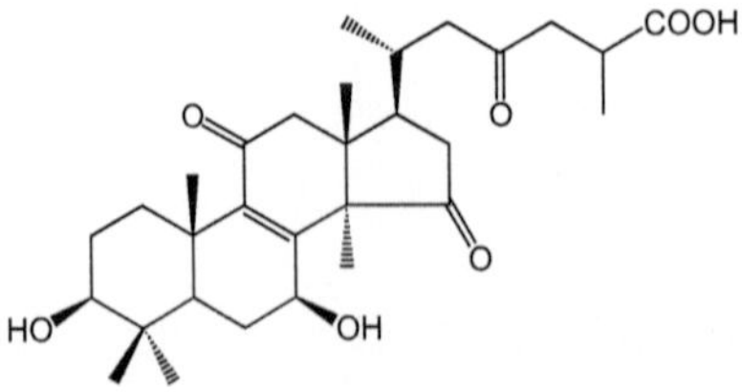

GA-B

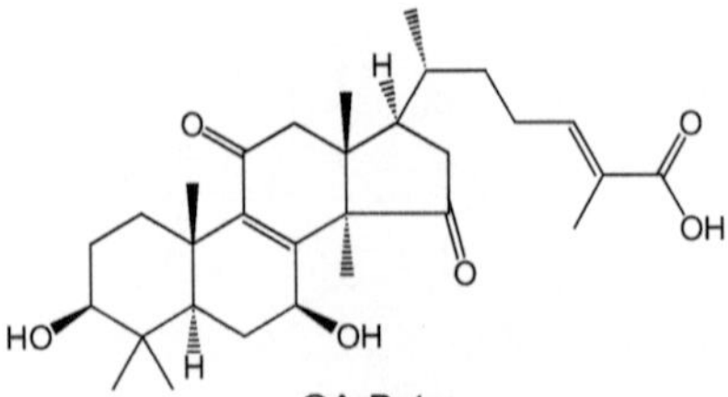

GA-Beta

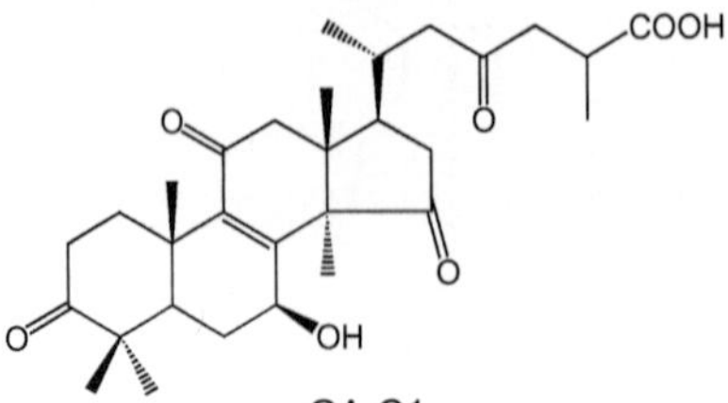

GA-C1

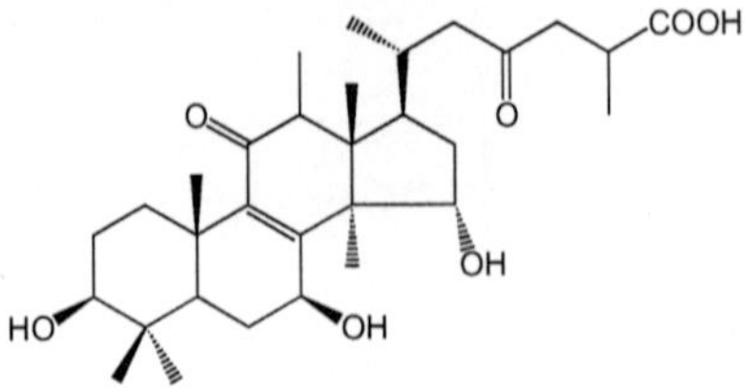

GA-C2

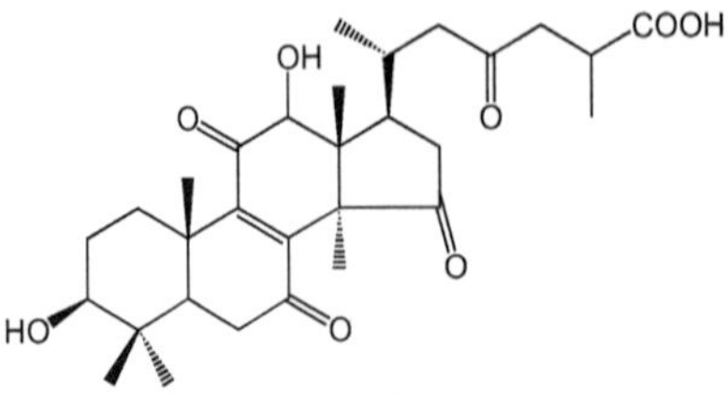

GA-C6

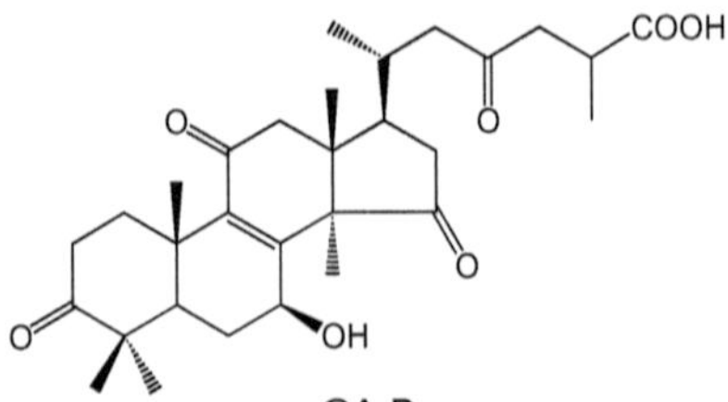

GA-D

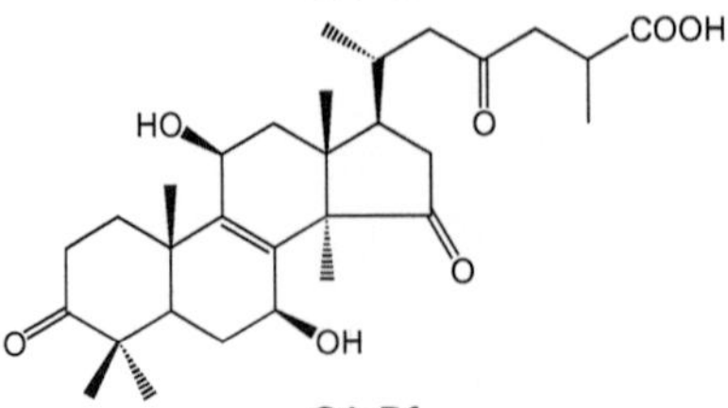

GA-Df

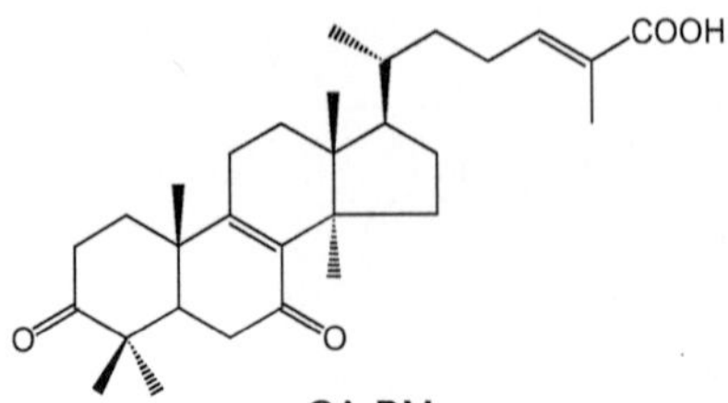

GA-DM

Figure 3.2—Cont'd

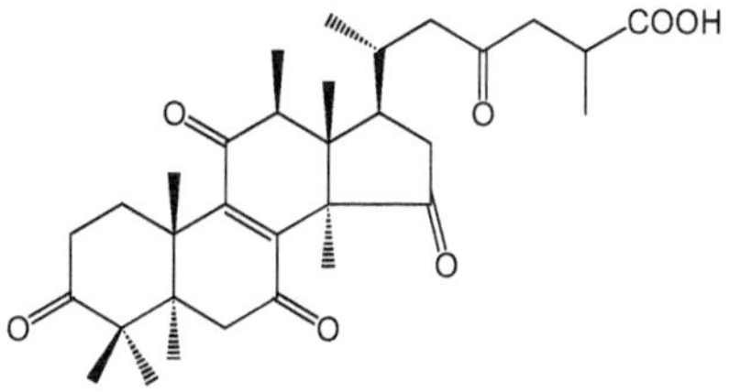

GA-E

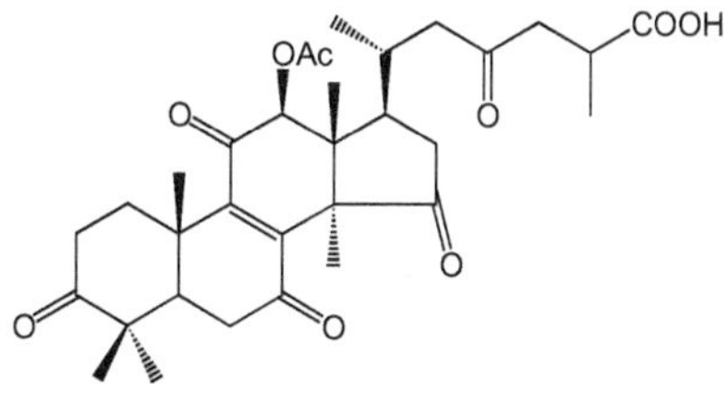

GA-F

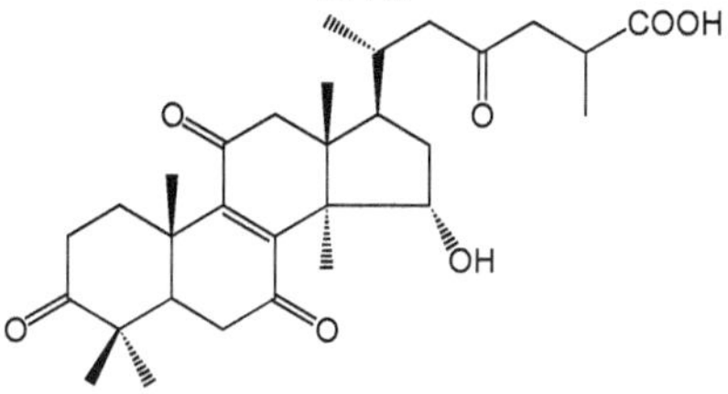

GA-G

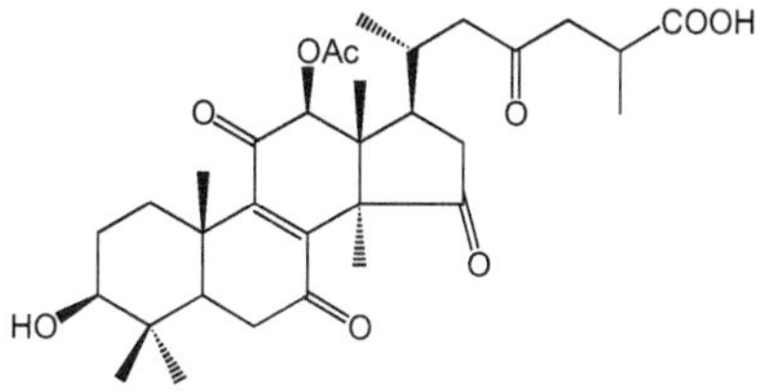

GA-H

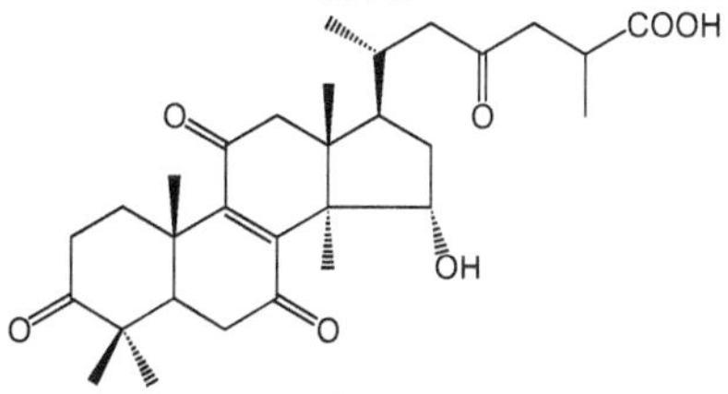

GA-J

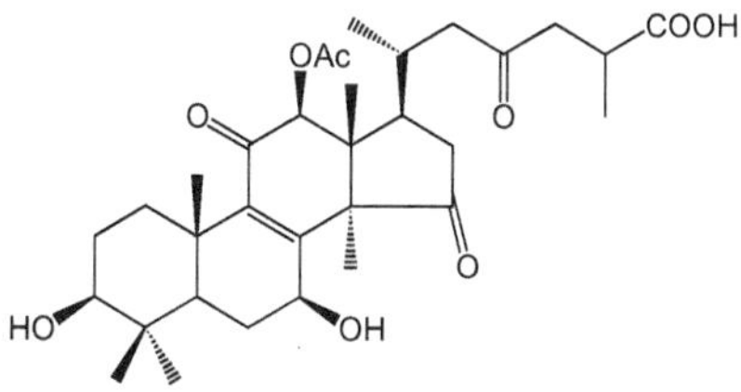

GA-K

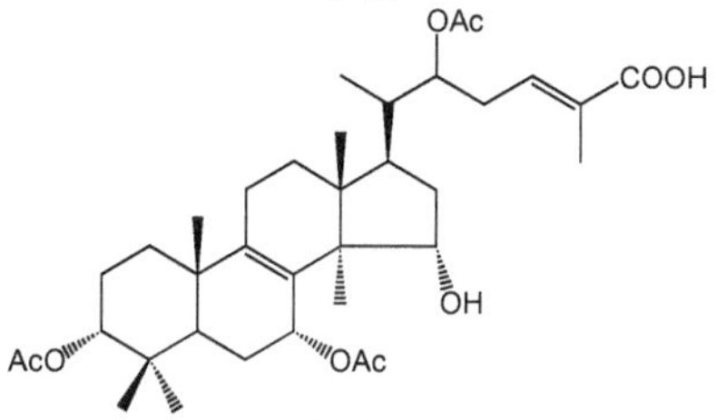

GA-Mc

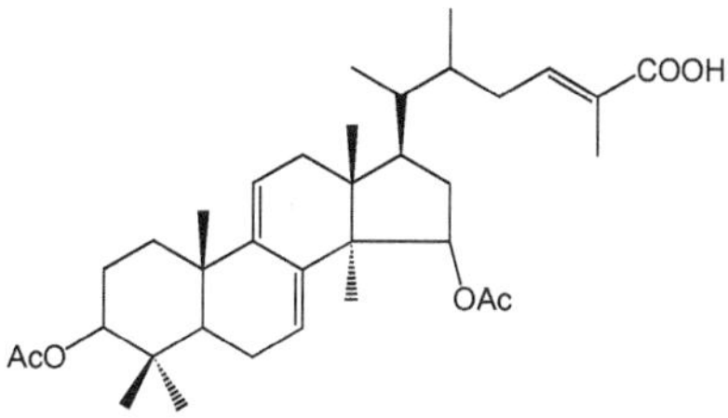

GA-Me

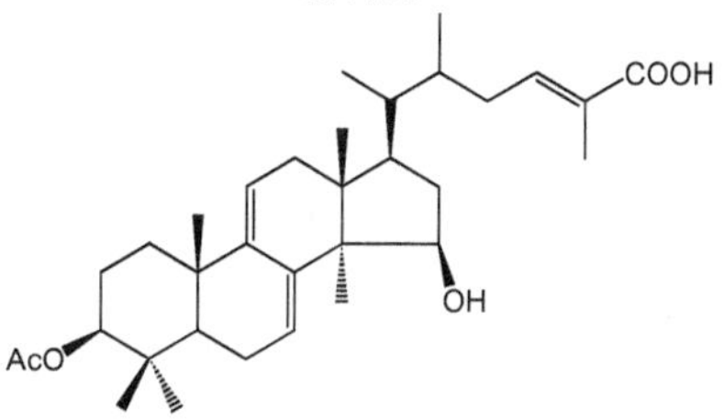

GA-Mf

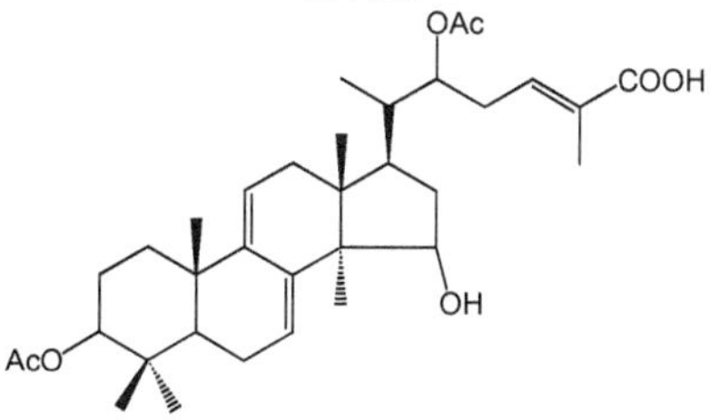

GA-Mk

Figure 3.2—Cont'd

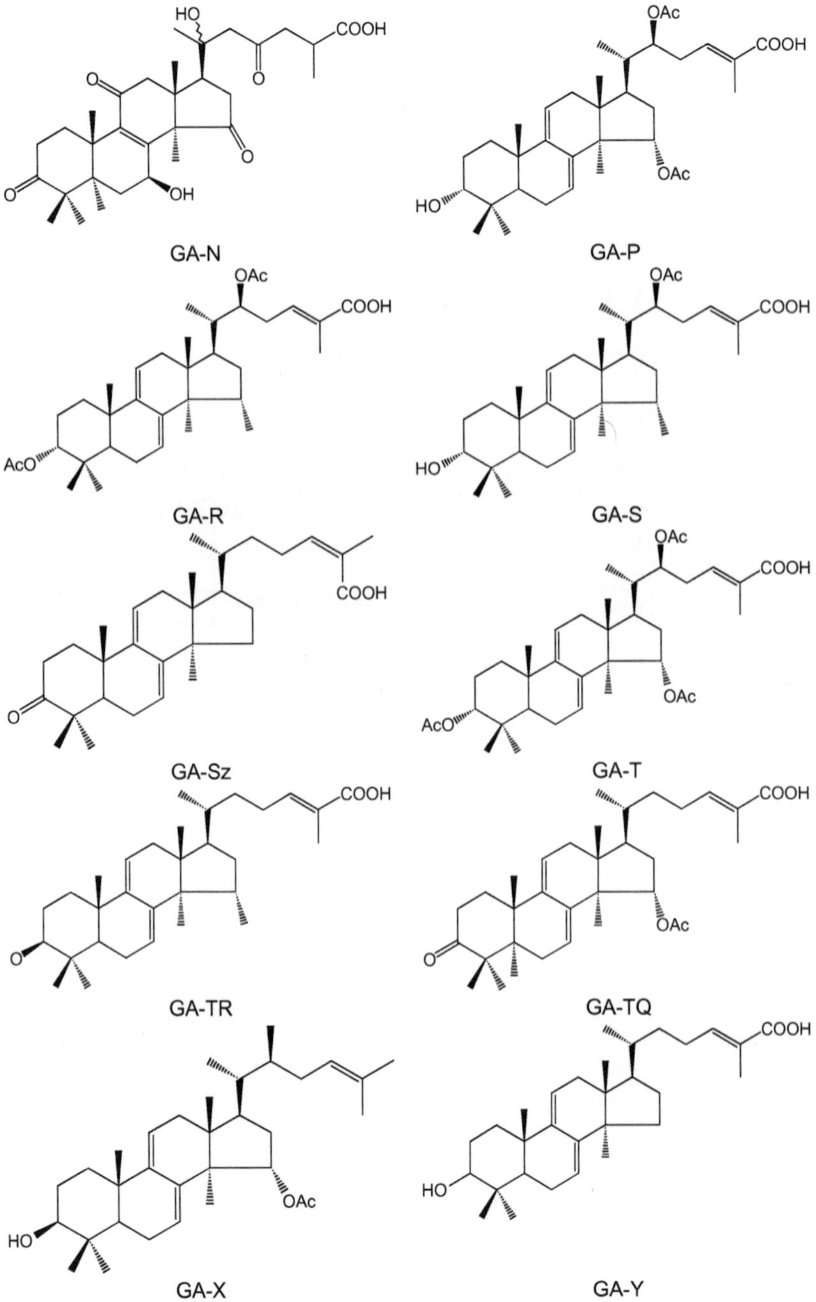

Figure 3.2 Chemical structures of ganoderic acids (GAs).

synthase (*SQS*), lanosterol synthase (*LS*), and methyl jasmonate (*MeJA*) genes are involved in ganoderic acid biosynthetic pathways in *G. lucidum*. *HMGR* catalyzes the first step in mevalonate pathway of isoprenoid biosynthesis [65]. Lichtenthaler *et al.* [66] indicated that *HMGR* plays important roles in mevalonate synthesis and key enzyme in ganoderic acids' biosynthesis. High level of *HMGR* gene will be able to the upregulation of mevalonate pathway and enhancing ganoderic acid production in the basidiomycete *G. lucidum* [48,67]. *SQS* had been proposed to be involved in the first enzymatic step from the central isoprenoid pathway of sterol and triterpenoid biosynthesis [65]. The over expression of *SQS* transcription factor led to increased production of ganoderic acid in liquid static culture of *G. lucidum* [68]. Shiao [69] indicated that *LS* catalyzes the formation of lanosterol in the early stages of ganoderic acid biosynthesis. *MeJA* is involved in the modulation of the gene levels of *HMGS, HMGR, MVD, FPS, SQS,* and *LANO* in the ganoderic acid biosynthesis pathway [70]. The biosynthesis of ganoderic acid from final step of lanosterol formation includes series actions of oxidation, reduction, hydroxylation, and acetylation [62]. Liang *et al.* [71] indicated that the phenobarbital enhanced biosynthetic gene expressions of *HMGR, SQS,* and *LS* and production of ganoderic acids in static liquid culture of *G. lucidum*. You *et al.* [72] indicated that enhancing ganoderic acid production is linked to the modulations of fungal apoptosis-signaling cascade and triterpenoid biosynthesis pathway. This study also indicated that aspirin was able to induce fungal apoptosis in *G. lucidum*, ROS accumulation, and phosphorylation of Hog-1 kinase in ganoderic acid biosynthetic regulation. Zhang *et al.* [73] indicated that higher gaseous oxygen level stimulated H_2O_2 production in *G. lucidum* and might be involved in the regulation of ganoderic acid biosynthesis. This report also indicated that a hydrogen peroxide (H_2O_2) release plays an important role in ganoderic acid metabolism in *G. lucidum* through induction of the transcription of *HMGR, SQS,* and *LS* genes.

2.3. Optimization of the Fermentation Process

The commercialization of bioactive compounds (such as polysaccharides and triterpenoids) from *Ganoderma* species was limited by low yields and high costs. Moreover, the contents of polysaccharides, lucidenic acids, and ganoderic acids isolated from fruit body of *Ganoderma* species are too low and very expensive. The mycelia cultures of *Ganoderma* species have been known to have the higher yields of bioactive constituents and lower costs than fruit body of *Ganoderma* species. Wei *et al.* [51] indicated that optimization of the fermentation medium and best of *Ganoderma* strain screening play key roles in the production of bioactive constituents (Table 3.1). This report also indicated that optimization of the fermentation medium (including 55 g/L of glucose, 14 g/L of yeast extract, 0.3 g/L of ferric acid, and other medium components unchanged) in the 10-L bioreactor enhanced the percentage of mycelia biomass and contents of total ganoderic acids and ganoderic acid T were 85%, 49%, and 93%, respectively. Moreover, optimization of the fermentation medium for the biomass and production of ganoderic acids was 23.9 g/L and 670 mg/L in 300-L bioreactor, respectively. Zhao *et al.* [44] indicated that a maximal content of ganoderic acids Mk, T, S, and Me was 2.16, 11.76, 31.09, and 7.04 μg/mg cell dry weight (DW), respectively, in static liquid culture of *G. lucidum* under the optimization of nitrogen-limiting condition (3 mM glutamine). You *et al.* [45] indicated that solid potato dextrose medium culture strategy resulted in the enhanced production of ganoderic acids in *G. lucidum* via the regulation of *SQS* and *LS* genes. Liu *et al.* [46] indicated that a maximal yield of ganoderic acid Me was 11.9 mg/L in submerged culture of *G. lucidum* using response surface methodology (optimization of glucose, peptone, and culture time was 44.4 g/L, 5.0 g/L, and 437.1 h, respectively). Tang *et al.* [43] indicated that a maximal production of ganoderic acids was 754.6 mg/L at a pH-shift and dissolved oxygen tension-shift integrated fed-batch fermentation process. Xu and Zhong [49] indicated that the addition of calcium ion (at 10 mM) to static liquid cultures of *G. lucidum* was able to induce the ganoderic acid biosynthesis through calmodulin-dependent calcineurin signal pathway. Xu *et al.* [50] also indicated that sodium ion (at 100 mM) could induce the ganoderic acid biosynthesis in static liquid

Table 3.1 Optimization of the Fermentation Process

Enhanced Production of Ganoderic Acids	**References**
Best of *Ganoderma* strain screening	[51]
Optimization of the fermentation medium (including nutrition factor and special medium)	[44–46,51]
A pH-shift and dissolved oxygen tension-shift integrated fed-batch fermentation process	[43]
Addition of calcium ion and sodium ion in static liquid culture	[49,50]
Static culture of *Ganoderma lucidum*	[47]

culture of *G. lucidum* through calcineurin signal transduction. Xu *et al.* [47] indicated that the contents of total crude ganoderic acids and ganoderic acids T, Me, Mk, and S in static culture of *G. lucidum* were higher than liquid-shaking culture. The contents of total crude ganoderic acids and ganoderic acids T, Me, Mk, and S in shaking culture were 1.1 mg/100 mg DW, 19.2 μg/100 mg DW, 2.1 μg/100 mg DW, 6.8 μg/100 mg DW, and 1.6 μg/100 mg DW, respectively. The contents of total crude ganoderic acids and ganoderic acids T, Me, Mk, and S in static culture were 2.3 mg/100 mg DW, 98.1 μg/100 mg DW, 29.2 μg/100 mg DW, 28.5 μg/100 mg DW, and 28.0 μg/100 mg DW, respectively.

3. BIOLOGICAL FUNCTIONS OF LUCIDENIC ACIDS AND GANODERIC ACIDS

3.1. Cytotoxic and Apoptotic Effects

G. lucidum is known as a functional mushroom used in many Asian countries, such as Taiwan, China, and Japan. Triterpenoids were isolated from *Ganoderma* species, which have been reported regarding the inhibition of tumor growth, induced cell apoptosis, regulated cell cycle arrest, and autophagy [6,10,38,39,74–76]. Hsu *et al.* [10] indicated that the cytotoxic and apoptotic effects of lucidenic acids A, B, C, and N from a new *G. lucidum* (YK-02) strain on HL-60 human leukemia cells might be through the mitochondria pathway. This report also revealed that the lucidenic acid B leads to the release of mitochondria cytochrome *c*, the regulation of Bcl-2 family members, the activations of caspase-9 and caspase-3, and then the cleavage of poly(ADP-ribose) polymerase. Jiang *et al.* [75] reported that lanostane-type triterpenes-ganoderic acids A and H inhibited cell growth in MDA-MB-231 human breast cancer cells.

Table 3.2 shows cytotoxicity effects of lucidenic acids on various cancer cell lines. Hsu *et al.* [10] indicated that the inhibitory effects of lucidenic acids A, B, C, and N on cell growth in human acute promyelocytic leukemia cells (HL-60), human hepatoblastoma cells (HepG2), and human colorectal carcinoma cells (HCT-116, COLO205, and HT-29) were determined by MTT assay. Lucidenic acids A, B, C, and N have the strongest inhibitory effect on HL-60 cell growth among the five human cancer cell lines and did not affect the cell growth of normal peripheral blood lymphocytes. Moreover, the IC_{50} (inhibits growth of 50%) values of lucidenic acids B and C on HL-60 cells were 19.3 and 45.0 μM, respectively. Wu *et al.* [77] indicated that the cytotoxicity effects of lucidenic acids A and

Table 3.2 Cytotoxicity Effects of Lucidenic Acids (LAs) on Various Cancer Cell Lines

Lucidenic Acids	Incubation Time (h)	Cancer Cell Lines	IC_{50}	References
LA-A	–	HepG2	0.164 nM	[77]
LA-A	–	HepG2,2,15	0.205 nM	[77]
LA-A	–	KB	16.97 μM	[77]
LA-A	–	CCM2	27.51 μM	[77]
LA-A	–	P388	0.017 μM	[77]
LA-A	72	COLO205	154 μM	[10]
LA-A	72	HCT-116	428 μM	[10]
LA-A	72	HepG2	183 μM	[10]
LA-A	72	HL-60	61.0 μM	[10]
LA-B	72	COLO205	249 μM	[10]
LA-B	72	HepG2	112 μM	[10]
LA-B	72	HL-60	19.3 μM	[10]
LA-B	72	HT-29	382 μM	[10]
LA-C	72	COLO205	195 μM	[10]
LA-C	72	HepG2	193 μM	[10]
LA-C	72	HL-60	45.0 μM	[10]
LA-N	72	COLO205	486 μM	[10]
LA-N	72	HepG2	230 μM	[10]
LA-N	72	HL-60	64.5 μM	[10]
LA-N	–	HepG2	0.206 nM	[77]
LA-N	–	HepG2,2,15	1.66 nM	[77]
LA-N	–	KB	26.69 μM	[77]
LA-N	–	CCM2	35.49 μM	[77]
LA-N	–	P388	0.012 μM	[77]

–, not provided.

N from the fruiting bodies of *G. lucidum* on murine leukemia cells (P388), cerebral cavernous malformations (CCM2), human hepatoblastoma cells (HepG2 and HepG2,2,15), and human oral cancer cells (KB) were examined. Lucidenic acid A showed the strongest inhibitory effect on human hepatoblastoma cell growth among the different types of cancer cell lines.

The IC_{50} values of lucidenic acids A and N on the different types of cancer cell lines are summarized in Table 3.3.

3.2. Cell Cycle Arrest

Inhibition of cancer cell population growth might be mediated through cell cytotoxic effect and cell cycle arrest. Many literatures have indicated that cell growth inhibition, cell cycle distribution, and cell apoptosis were evaluated using flow cytometry by the propidium iodide staining methods. Zhu *et al.* [81] indicated that alcohol extracts from the spores of *G. lucidum* led to cell cycle arrest at the transition from G1 to S phase. Hu *et al.* [82] indicated that ganoderic acid from the fruiting body induced cell cycle arrest at the G1 phase in HeLa human cervical carcinoma cells. Yang [74] indicated that ganoderic acid from submerged culture of *G. lucidum* (500 μg/mL, 12–60 h) led to cell cycle arrest at the G1 phase in human hepatoma cell line BEL7402. Yue *et al.* [83] indicated that ganoderic acid D isolated from *G. lucidum* induced cell cycle arrest at the G2/M phase and cell apoptosis in HeLa human cervical carcinoma cells. Wu *et al.* [31] indicated that ganoderic acid DM caused cell cycle arrest in the G1 phase in MCF-7 human breast cancer cells through modulation of CDK2, CDK6, cycle D1, p-Rb, and c-Myc expressions.

3.3. Anti-invasive Effect

G. lucidum is a well-known mushroom with various biological effects and used for preventing or treating human diseases such as arthritis, allergy, bronchitis, chronic hepatitis, hyperglycemia, hepatopathy, inflammation, nephritis, and neurasthenia [84–86]. Many reports indicated that the possible mechanisms for the anticancer effects of lucidenic acids and ganoderic acids from *G. lucidum* involved in direct cytotoxicity and apoptosis, anti-angiogenesis, anti-invasion, and antimetastasis [18,82,87,88]. Cancer metastasis is a major cause of death in patients with a variety of cancers and involves degradation of environmental barriers [89]. The degradation of environmental barriers such as extracellular matrix (ECM) and basement membrane plays an important role in the invasion and metastasis of various cancer cells. Secretions of the matrix metalloproteinases (MMP)-2 and MMP-9 are highly expressed in many malignant tumors, such as lung, colon, and breast cancers [90]. MMP-2/-9 plays important roles in cancer cell invasion, metastasis, and angiogenesis and can degrade multiple substrates, such as ECM [91]. Numerous lucidenic acids (including A, B, C, and N) were isolated from the fruiting bodies of new *G. lucidum* strain (YK-02) and

Table 3.3 Cytotoxicity Effects of Ganoderic Acids (GAs) on Various Cancer Cell Lines

Ganoderic Acids	Incubation Time (h)	Cancer Cell Lines	IC_{50}	References
GA-A	48	LLC	>20 μg/mL	[15]
GA-A	48	T-47D	>20 μg/mL	[15]
GA-A	48	S-180	>20 μg/mL	[15]
GA-A	48	Meth-A	>20 μg/mL	[15]
GA-AM1	48	HeLa	19.8 μM	[13]
GA-B	48	HeLa	20.3 μM	[13]
GA-B8	48	LLC	>20 μg/mL	[15]
GA-B8	48	T-47D	>20 μg/mL	[15]
GA-B8	48	S-180	>20 μg/mL	[15]
GA-B8	48	Meth-A	>20 μg/mL	[15]
GA-C1	48	LLC	>20 μg/mL	[15]
GA-C1	48	T-47D	>20 μg/mL	[15]
GA-C1	48	S-180	>20 μg/mL	[15]
GA-C1	48	Meth-A	>20 μg/mL	[15]
GA-D	48	HeLa	17.3 μM	[13]
GA-DM	72	PC-3	40.0 μM	[46]
GA-E	–	HepG2	0.144 nM	[77]
GA-E	–	HepG2,2,15	0.105 nM	[77]
GA-E	–	CCM2	31.25 μM	[77]
GA-E	–	P388	5.012 μM	[77]
GA-F	48	HeLa	19.5 μM	[13]
GA-K	48	HeLa	15.1 μM	[13]
GA-Mc	–	HeLa	35.4 μM	[24]
GA-Mc	–	95-D	24.0 μM	[24]
GA-Mf	24	LO_2	77.0 μM	[78]
GA-Mf	24	HF	122.5 μM	[78]
GA-Mf	24	HO-8910PM	34.8 μM	[78]

Table 3.3 Cytotoxicity Effects of Ganoderic Acids (GAs) on Various Cancer Cell Lines—cont'd

Ganoderic Acids	Incubation Time (h)	Cancer Cell Lines	IC_{50}	References
GA-Mf	24	SW1990	47.1 μM	[78]
GA-Mf	24	95-D	33.2 μM	[78]
GA-Mf	24	HeLa	36.4 μM	[78]
GA-Mf	–	HeLa	36.4 μM	[24]
GA-Mf	–	95-D	33.2 μM	[24]
GA-Mk	–	HeLa	29.8 μM	[24]
GA-Mk	–	95-D	30.8 μM	[24]
GA-P	–	NCI-H187	26.0 μM	[59]
GA-P	–	MCF-7	58.0 μM	[59]
GA-P	–	KB	40.0 μM	[59]
GA-R	–	HeLa	21.4 μM	[24]
GA-R	–	95-D	31.9 μM	[24]
GA-S	24	LO_2	44.7 μM	[78]
GA-S	24	HF	28.7 μM	[78]
GA-S	24	HO-8910PM	25.2 μM	[78]
GA-S	24	SW1990	41.4 μM	[78]
GA-S	24	95-D	38.5 μM	[78]
GA-S	24	HeLa	23.8 μM	[78]
GA-S	–	NCI-H187	39.0 μM	[59]
GA-S	–	MCF-7	>98 μM	[59]
GA-S	–	KB	53.0 μM	[59]
GA-S	–	HeLa	23.8 μM	[24]
GA-S	–	95-D	38.5 μM	[24]
GA-T	–	NCI-H187	15.0 μM	[59]
GA-T	–	MCF-7	78.0 μM	[59]
GA-T	–	KB	18.0 μM	[59]
GA-T	24	95-D	27.9 μg/mL	[79]

Continued

Table 3.3 Cytotoxicity Effects of Ganoderic Acids (GAs) on Various Cancer Cell Lines—cont'd

Ganoderic Acids	Incubation Time (h)	Cancer Cell Lines	IC_{50}	References
GA-T	24	HeLa	11.0 μM	[46]
GA-T	24	HepG2	13.1 μM	[46]
GA-T	24	95-D	14.5 μM	[46]
GA-X	24	HuH-7	20.3 μg/mL	[80]
GA-X	24	HCT-116	38.3 μg/mL	[80]
GA-X	24	Raji	39.2 μg/mL	[80]
GA-X	24	HL-60	26.5 μg/mL	[80]

–, not provided.

G. colossum strain, and all of them were found to have potential anti-invasive and antimetastatic activities [18,28,56,92].

In vitro studies, lucidenic acids A, B, C, and N isolated from new *G. lucidum* strain (YK-02) have anti-invasive activity in phorbol-12-myristate-13-acetate (PMA)-induced HepG2 human hepatoma cells [18]. Moreover, lucidenic acid B had the potentials to prevent PMA-induced invasion of HepG2 cells and inhibited MMP-9 activity through inhibitions of pERK1/2, AP-1, and NF-κB DNA-binding activity [56].

In vivo study, lucidenic acid-rich *G. lucidum* extract could effectively demonstrate the inhibitions of cell proliferation and metastasis in HepG2 human hepatoma cells implanted into ICR-*nu/nu* mice [92]. In other *Ganoderma* species, ethanol extract from the fruiting body of a *G. colossum* strain (EEGC) is an effective inhibitor on the PMA-induced invasion of HepG2 cells. Thus, the downregulations of pERK1/2, pAkt, AP-1, and NF-κB might be involved in inhibition of EEGC on the MMP-9 activity in PMA-induced HepG2 cells [28]. GDNT, a *G. lucidum* alcohol, inhibited invasive behavior (including cell adhesion, migration, and invasion) of MDA-MB-231 breast cancer cells through the downregulations of CDC20 and uPA [29].

3.4. Autophagy

Autophagy (also called self-eating) has been widely studied in different types of cancer therapy. Three main types of autophagy (including macroautophagy, microautophagy, and chaperon-mediated autophagy) have been reported in mammalian cells. The activities of autophagy-related (ATG) proteins (including ATG1, ATG5, ATG8, ATG12, ATG13, and ATG17) play a key role in autophagy. Hossain *et al.* [30] indicated that ganoderic acid DM from the medicinal mushroom *G. lucidum* induced autophagic and apoptotic cell death in melanoma cells through enhanced immunological responses (such as increased HLA class II presentation) and initiated a possible cross-talk between autophagy and apoptosis (such as upregulated protein expressions of Beclin 1 and LC3).

3.5. Anti-inflammatory Effect

Inflammatory response plays an important role in the immune system. Chronic inflammatory responses occur in many disorders, such as atherosclerosis, asthma, psoriasis, rheumatoid arthritis, inflammatory bowel disease, and cancer [93–96]. Ko *et al.* [26] indicated the anti-inflammatory properties of triterpenoids and steroids, which were isolated from both *G. lucidum* and *Ganoderma tsugae.* Twelve compounds isolated from both *Ganoderma* species, including ergosta-7,22-dien-3β-ol, ergosta-7,22-dien-3β-yl palmitate, ergosta-7,22-dien-3-one, ergosta-7,22-dien-2β,3α,9α-triol, 5α,8α-epidioxyergosta-6,22-dien-3β-ol, ganoderal A, ganoderal B, ganoderic aldehyde A, tsugaric acid A, 3-oxo-5α-lanosta-8,24-dien-21-oic acid, 3α-acetoxy-5α-lanosta-8,24-dien-21-oic acid ester β-D-glucoside, and tsugaric acid B. Moreover, tsugaric acid A from *G. tsugae* suppressed superoxide anion formation in fMLP/CB-stimulated rat neutrophils as well as protected ultraviolet B (UVB) light-induced damage in human keratinocytes. 3-Oxo-5α-lanosta-8,24-dien-21-oic acid isolated from *G. tsugae* could inhibit lipopolysaccharide (LPS)/interferon-γ-induced nitric oxide production in N9 microglial cells.

3.6. Antiosteoclastogenesis

Ganoderic acid DM isolated from the ethanol extracts of *G. lucidum* could block osteoclastogenesis on ovariectomized-induced deterioration of bone density in 11-week-old female Sprague-Dawley rats by suppressing the expressions of c-Fos and NFATc1 [27]. Liu *et al.* [97] also indicated that ganoderic acid DM from the ethanol extracts of *G. lucidum* caused growth inhibition in prostate cancer cells and suppressed osteoclastogenesis.

3.7. Antiasthma

Asthma is a chronic lung disease involving recurrent breathing problems. Eosinophil-predominant, neutrophil-predominant, and mixed eosinophil and neutrophil airway inflammation are regarded as the typical characteristics of asthma. Srivastava *et al.* [32] indicated that antiasthma simplified herbal medicine intervention prevents neutrophil-predominant airway inflammation in a ragweed-sensitized murine asthma model. Moreover, ganoderic acid C1 inhibits ragweed-induced TNF-α production in RAW 264.7 macrophages through decreased phosphorylated IκB expression and increased HDAC2 expression.

3.8. Antihepatitis B Activity

In antihepatitis B activity of ganoderic acid, Li and Wang [25] indicated that ganoderic acid (at 8 μg/mL, 8 days) from the cultivation of *G. lucidum* suppressed replication of hepatitis B virus (HBV) in HepG2,2,15 cells (HepG2 cell transferred with HBV DNA). Moreover, this report also revealed that ganoderic acid prevented *M. bovis* BCG plus LPS-induced liver injury in male KM mice.

4. PHARMACOKINETICS OF GANODERIC ACIDS

Pharmacokinetics/compound metabolism plays important role in identify and characterize the drug or natural compound of metabolites. High-performance liquid chromatography with UV detection (HPLC-UV) is the most widely used method in the past for the analysis of triterpenoids in *Ganoderma* species [2,98,99]. But, the chemical structures of a relatively larger number of triterpenoids are very similar; they seem to have resulted in the process of peak overlapping that occurs in HPLC-UV. However, rapid and specific methods for the quantification of natural compound metabolites in animals' (such as mice or rat) sample (such as plasma, urine, bile, or organ tissues) and human after oral administration were developed by using high-performance liquid chromatography coupled with diode array detection and electrospray ion trap tandem mass spectrometry (HPLC-DAD-ESI-MS) and liquid chromatography coupled with electrospray ionization hybrid ion trap and time-of-flight mass spectrometry (LC-ESI-IT-TOF/MS). Many reports have indicated that the metabolites and pharmacokinetics of ganoderic acids (such as ganoderic acid A, B, C2, D, and F) from *Ganoderma* genus *in vitro* and *in vivo* were investigated [52–55,100,101]. Cheng *et al.* [100] indicated that the major metabolite structures of ganoderic acid D were the 3-carbonyl group, angular methyl groups, the 7-hydroxy group, and the 26-carboxylic acid moiety in rat bile. Teekachunhatean *et al.* [101] indicated that the C_{max} (maximum concentration observed) and the T_{max} (time of maximum concentration observed) of plasma ganoderic acid A were 10.99 ng/mL and 0.54 h, respectively, after oral administration of 3000 mg Lingzhi preparation (including 1417.8 μg/g ganoderic acid A and 224.15 μg/g ganoderic acid F) in healthy male volunteers under a fasting condition. Moreover, the C_{max} and the T_{max} of plasma ganoderic acid F were 2.57 ng/mL and 0.52 h, respectively, after a single oral administration under a fasting condition. Fourteen metabolites of ganoderic acid B in rat plasma, bile, kidney, and stomach were detected by LC-ESI-IT-TOF/MS after oral administration of ganoderic acid B [52]. Guo *et al.* [53] indicated that the T_{max} of plasma ganoderic acid C2 was 3.9 min after an oral administration of ganoderic acid C2 at the dose of 20 mg/kg to male rats. Cheng *et al.* [54] indicated that pharmacokinetics of ganoderic acid D in plasma after oral administration of ganoderic acid D-loaded solid lipid nanoparticles and suspension and intravenous administration of ganoderic acid D solution was investigated by liquid chromatography-tandem mass spectrometry. The C_{max} and the T_{max} of ganoderic acid D in plasma after oral administration of ganoderic acid D-loaded solid lipid nanoparticles were 1555.59 ng/mL and 0.36 h, respectively. The C_{max} and the T_{max} of ganoderic acid D in plasma after oral administration of ganoderic acid D suspension were 107.18 ng/mL and 2.00 h, respectively. The C_{max} of ganoderic acid D in plasma was 3582.93 ng/mL, after intravenous administration of ganoderic acid D solution. Cheng *et al.* [55] indicated that the major metabolic soft spots of ganoderic acids from crude extract of *G. lucidum* in rats were 3,7,15,23-carbonyl groups (or hydroxyl groups), angular methyl groups, 20(22)-double bond, 12-acetoxyl group, and 26-carboxylic acid moiety. These reports are very helpful in the further study of the pharmacokinetics and clinical studies of bioactive triterpenoids (such as ganoderic acids) from *Ganoderma* species.

5. CONCLUSION

Finally, this chapter summarizes the studies describing the sources, biosynthesis, biological functions (including cell cytotoxicity, cell apoptosis, cell cycle arrest, anti-invasion, autophagy, anti-inflammation, anti-osteoclastogenesis, antiasthma, and antihepatitis B activity), and pharmacokinetics of lucidenic acids and ganoderic acids. The available scientific evidence indicates that ganoderic acids and lucidenic acids from *Ganoderma* species may provide a therapeutic approach for delaying or preventing multiple human diseases.

REFERENCES

[1] S. Kimura, T. Tamura, Dietary effect of *Ganoderma lucidum* mushroom on blood pressure and lipid levels in spontaneously hypertensive rats (SHR), J. Nutr. Sci. Vitaminol. 34 (1988) 433–438.

[2] B.S. Min, et al., Triterpenes from the spores of *Ganoderma lucidum* and their inhibitory activity against HIV-1 protease, Chem. Pharm. Bull. 46 (1998) 1607–1612.

[3] D.H. Kim, et al., Beta-glucuronidase-inhibitory activity and hepatoprotective effect of *Ganoderma lucidum*, Biol. Pharm. Bull. 22 (1999) 162–164.

[4] C.Y. Su, M.S. Shiao, C.T. Wang, Differential effects of ganodermic acid S on the thromboxane A2-signaling pathways in human platelets, Biochem. Pharmacol. 58 (1999) 587–595.

[5] M. Zhu, et al., Triterpene antioxidants from *Ganoderma lucidum*, Phytother. Res. 13 (1999) 529–531.

[6] S.B. Lin, et al., Triterpene-enriched extracts from *Ganoderma lucidum* inhibit growth of hepatoma cells via suppressing protein kinase C, activating mitogen-activated protein kinases and G2-phase cell cycle arrest, Life Sci. 72 (2003) 2381–2390.

[7] H. Hajjaj, et al., Effect of 26-oxygenosterols from *Ganoderma lucidum* and their activity as cholesterol synthesis inhibitors, Appl. Environ. Microbiol. 71 (2005) 3653–3658.

[8] L.N. Ofodile, N.U. Uma, M.S.J. Simmonds, Antimicrobial colossolactones from a Nigerian polypore *Ganoderma colossum* (Fr.) C.F. Baker, Int. J. Med. Mushrooms 7 (2005) 437–438.

[9] H.J. Timo, U. Lindequist, L. Michael, Antivital terpenoid constituents of *Ganoderma pfeifferi*, J. Nat. Prod. 68 (2005) 1728–1731.

[10] C.L. Hsu, Y.S. Yu, G.C. Yen, Lucidenic acid B induces apoptosis in human leukemia cells via a mitochondria-mediated pathway, J. Agric. Food Chem. 56 (11) (2008) 3973–3980.

[11] S.C. Hsu, C.C. Ou, M.C. Kao, *Ganoderma tsugae* extracts inhibit colorectal cancer cell growth via G2/M cell cycle arrest, J. Ethnopharmacol. 120 (2008) 394–401.

[12] J. Liu, et al., Ganoderic acids from *Ganoderma* lucidum: Inhibitory activity of osteoclastic differentiation and structural criteria, Planta Med. 76 (2010) 137–139.

[13] Q.X. Yue, et al., Effects of triterpenes from *Ganoderma lucidum* on protein expression profile of HeLa cells, Phytomedicine 17 (2010) 606–613.

[14] Y. Mizushina, et al., Lucidenic acid O and lactone, new terpene inhibitors of eukaryotic DNA polymerases from a basidiomycete, *Ganoderma lucidum*, Bioorg. Med. Chem. 7 (9) (1999) 2047–2052.

[15] J.J. Gao, et al., New triterpene aldehydes, lucialdehydes A–C, from *Ganoderma lucidum* and their cytotoxicity against murine and human tumor cells, Chem. Pharm. Bull. 50 (6) (2002) 837–840.

[16] K. Iwatsuki, et al., Lucidenic acids P and Q, methyl lucidenate P, and other triterpenoids from the fungus *Ganoderma lucidum* and their inhibitory effects on Epstein–Barr virus activation, J. Nat. Prod. 66 (12) (2003) 1582–1585.

[17] J. Liu, et al., 5alpha-reductase inhibitory effect of triterpenoids isolated from *Ganoderma lucidum*, Biol. Pharm. Bull. 29 (2) (2006) 392–395.

[18] C.J. Weng, et al., The anti-invasive effect of lucidenic acids isolated from a new *Ganoderma lucidum* strain, Mol. Nutr. Food Res. 51 (12) (2007) 1472–1477.

[19] T. Akihisa, et al., Anti-inflammatory and anti-tumor-promoting effects of triterpene acids and sterols from the fungus *Ganoderma lucidum*, Chem. Biodivers. 4 (2) (2007) 224–231.

[20] H.W. Seo, et al., Steroids and triterpenes from the fruit bodies of *Ganoderma lucidum* and their anti-complement activity, Arch. Pharm. Res. 32 (11) (2009) 1573–1579.

[21] I. Lee, et al., Selective cholinesterase inhibition by lanostane triterpenes from fruiting bodies of *Ganoderma lucidum*, Bioorg. Med. Chem. Lett. 21 (21) (2011) 6603–6607.

[22] U. Grienke, et al., Pharmacophore-based discovery of FXR-agonists. Part II: Identification of bioactive triterpenes from Ganoderma lucidum, Bioorg. Med. Chem. 19 (22) (2011) 6779–6791.
[23] N. Ding, et al., Separation and determination of four ganoderic acids from dried fermentation mycelia powder of *Ganoderma lucidum* by capillary zone electrophoresis, J. Pharm. Biomed. Anal. 53 (5) (2010) 1224–1230.
[24] Y.B. Li, R.M. Liu, J.J. Zhong, A new ganoderic acid from *Ganoderma lucidum* mycelia and its stability, Fitoterapia 84 (2013) 115–122.
[25] Y.Q. Li, S.F. Wang, Anti-hepatitis B activities of ganoderic acid from *Ganoderma lucidum*, Biotechnol. Lett. 28 (11) (2006) 837–841.
[26] H.H. Ko, et al., Antiinflammatory triterpenoids and steroids from *Ganoderma lucidum* and *G. tsugae*, Phytochemistry 69 (1) (2008) 234–239.
[27] I. Miyamoto, et al., Regulation of osteoclastogenesis by ganoderic acid DM isolated from *Ganoderma lucidum*, Eur. J. Pharmacol. 602 (1) (2009) 1–7.
[28] C.J. Weng, et al., Anti-invasive effect of a rare mushroom, *Ganoderma colossum*, on human hepatoma cells, J. Agric. Food Chem. 58 (13) (2010) 7657–7663.
[29] J. Jiang, A. Jedinak, D. Sliva, Ganodermanontriol (GDNT) exerts its effect on growth and invasiveness of breast cancer cells through the down-regulation of CDC20 and uPA, Biochem. Biophys. Res. Commun. 415 (2) (2011) 325–329.
[30] A. Hossain, et al., A possible cross-talk between autophagy and apoptosis in generating an immune response in melanoma, Apoptosis 17 (10) (2012) 1066–1078.
[31] G.S. Wu, et al., Ganoderic acid DM, a natural triterpenoid, induces DNA damage, G1 cell cycle arrest and apoptosis in human breast cancer cells, Fitoterapia 83 (2) (2012) 408–414.
[32] K.D. Srivastava, et al., Effect of antiasthma simplified herbal medicine intervention on neutrophil predominant airway inflammation in a ragweed sensitized murine asthma model, Ann. Allergy Asthma Immunol. 112 (4) (2014) 339–347.
[33] J.O. Toth, B. Luu, G. Ourisson, Les acides ganoderiques tàz: Triterpenes cytotoxiques de Ganoderma lucidum (Polyporacée), Tetrahedron Lett. 24 (1983) 1081–1084.
[34] H. Kohda, et al., The biologically active constituents of *Ganoderma lucidum* (Fr.) Karst: Histamine release-inhibitory triterpenes, Chem. Pharm. Bull. 33 (1985) 1367–1373.
[35] K. Tasaka, et al., Anti-allergic constituents in the culture medium of *Ganoderma lucidum* (II): The inhibitory effect of cyclooctasulfur on histamine release, Agents Actions 23 (1988) 157–160.
[36] H. Hikino, et al., Mechanisms of hypoglycemic activity of ganoderan B: A glycan of *Ganoderma lucidum* fruit bodies, Planta Med. 55 (5) (1989) 423–428.
[37] K. Kino, et al., Isolation and characterization of a new immunomodulatory protein, ling zhi-8 (LZ-8), from *Ganoderma lucidium*, J. Biol. Chem. 264 (1) (1989) 472–478.
[38] B.S. Min, et al., Triterpenes from the spores of *Ganoderma lucidum* and their cytotoxicity against meth-A and LLC tumor cells, Chem. Pharm. Bull. 48 (7) (2000) 1026–1033.
[39] Y. Kimura, M. Taniguchi, K. Baba, Antitumor and antimetastatic effects on liver of triterpenoid fractions of *Ganoderma lucidum*: Mechanism of action and isolation of an active substance, Anticancer Res 22 (6A) (2002) 3309–3318.
[40] D. Sliva, et al., *Ganoderma lucidum* suppresses motility of highly invasive breast and prostate cancer cells, Biochem. Biophys. Res. Commun. 298 (4) (2002) 603–612.
[41] G. Stanley, et al., *Ganoderma lucidum* suppresses angiogenesis through the inhibition of secretion of VEGF and TGF-beta1 from prostate cancer cells, Biochem. Biophys. Res. Commun. 330 (1) (2005) 46–52.
[42] J.J. Gao, et al., In vivo antitumor effects of bitter principles from the antlered form of fruiting bodies of *Ganoderma lucidum*, J. Nat. Med. 60 (2006) 42–48.
[43] Y.J. Tang, W. Zhang, J.J. Zhong, Performance analyses of a pH-shift and DOT-shift integrated fed-batch fermentation process for the production of ganoderic acid and

Ganoderma polysaccharides by medicinal mushroom *Ganoderma lucidum*, Bioresour. Technol. 100 (5) (2009) 1852–1859.
[44] W. Zhao, J.W. Xu, J.J. Zhong, Enhanced production of ganoderic acids in static liquid culture of *Ganoderma lucidum* under nitrogen-limiting conditions, Bioresour. Technol. 102 (17) (2011) 8185–8190.
[45] B.J. You, et al., Enhanced production of ganoderic acids and cytotoxicity of *Ganoderma lucidum* using solid-medium culture, Biosci. Biotechnol. Biochem. 76 (8) (2012) 1529–1534.
[46] G.Q. Liu, et al., Improving the fermentation production of the individual key triterpene ganoderic acid by the medicinal fungus *Ganoderma lucidum* in submerged culture, Molecules 17 (11) (2012) 12575–12586.
[47] J.W. Xu, et al., Isolation and analysis of differentially expressed genes during asexual sporulation in liquid static culture of *Ganoderma lucidum* by suppression subtractive hybridization, Mol. Biol. Rep. 39 (4) (2012) 3603–3610.
[48] J.W. Xu, Y.N. Xu, J.J. Zhong, Enhancement of ganoderic acid accumulation by overexpression of an N-terminally truncated 3-hydroxy-3-methylglutaryl coenzyme A reductase gene in the basidiomycete *Ganoderma lucidum*, Appl. Environ. Microbiol. 78 (22) (2012) 7968–7976.
[49] Y.N. Xu, J.J. Zhong, Impacts of calcium signal transduction on the fermentation production of antitumor ganoderic acids by medicinal mushroom *Ganoderma lucidum*, Biotechnol. Adv. 30 (6) (2012) 1301–1308.
[50] Y.N. Xu, X.X. Xia, J.J. Zhong, Induced effect of Na (+) on ganoderic acid biosynthesis in static liquid culture of *Ganoderma lucidum* via calcineurin signal transduction, Biotechnol. Bioeng. 110 (7) (2013) 1913–1923.
[51] Z.H. Wei, et al., Screening of *Ganoderma* strains with high polysaccharides and ganoderic acid contents and optimization of the fermentation medium by statistical methods, Bioprocess Biosyst, Eng. 37 (9) (2014) 1789–1797.
[52] X. Guo, et al., Structural identification of the metabolites of ganoderic acid B from *Ganoderma lucidum* in rats based on liquid chromatography coupled with electrospray ionization hybrid ion trap and time-of-flight mass spectrometry, Biomed. Chromatogr. 27 (9) (2013) 1177–1187.
[53] X.Y. Guo, et al., Structural characterization of minor metabolites and pharmacokinetics of ganoderic acid C2 in rat plasma by HPLC coupled with electrospray ionization tandem mass spectrometry, J. Pharm. Biomed. Anal. 75 (2013) 64–73.
[54] C.R. Cheng, et al., Pharmacokinetics of ganoderic acid D and its main metabolite by liquid chromatography-tandem mass spectrometry, J. Chromatogr. B Analyt. Technol. Biomed. Life Sci. 930 (2013) 1–6.
[55] C.R. Cheng, et al., Metabolite identification of crude extract from *Ganoderma lucidum* in rats using ultra-performance liquid chromatography-quadrupole time-of-flight mass spectrometry, J. Chromatogr. B Analyt. Technol. Biomed. Life Sci. 941 (2013) 90–99.
[56] C.J. Weng, et al., Lucidenic acid inhibits PMA-induced invasion of human hepatoma cells through inactivating MAPK/ERK signal transduction pathway and reducing binding activities of NF-kappaB and AP-1, Carcinogenesis 29 (10) (2008) 147–156.
[57] F. Wang, J.K. Liu, Highly oxygenated lanostane triterpenoids from the fungus *Ganoderma applanatum*, Chem. Pharm. Bull. 56 (7) (2008) 1035–1037.
[58] X.R. Peng, et al., Protective effects of triterpenoids from *Ganoderma resinaceum* on H_2O_2-induced toxicity in HepG2 cells, Food Chem. 141 (2) (2013) 920–926.
[59] M. Isaka, et al., Lanostane triterpenes from cultures of the Basidiomycete *Ganoderma orbiforme* BCC 22324, Phytochemistry 87 (2013) 133–139.
[60] S.R. El-Mekkaway, et al., Anti-HIV-1 and anti-HIV-protease substances from *Ganoderma lucidum*, Phytochemistry 49 (1998) 1651–1657.

[61] R.R.M. Paterson, Ganoderma—A therapeutic fungal biofactory, Phytochemistry 67 (2006) 1985–2001.
[62] J.W. Xu, W. Zhao, J.J. Zhong, Biotechnological production and application of ganoderic acids, Appl. Microbiol. Biotechnol. 87 (2010) 457–466.
[63] L. Shi, et al., Current progress in the study on biosynthesis and regulation of ganoderic acids, Appl. Microbiol. Biotechnol. 88 (6) (2010) 1243–1251.
[64] A. Ren, et al., Methyl jasmonate induces ganoderic acid biosynthesis in the basidiomycetous fungus *Ganoderma lucidum*, Bioresour. Technol. 101 (17) (2010) 6785–6790.
[65] I. Abe, M. Rohmer, G.D. Prestwich, Enzymatic cyclization of squalene and oxidosqualene to sterols and triterpenes, Chem. Rev. 93 (1993) 2189–2206.
[66] H.K. Lichtenthaler, M. Rohmer, J. Schwender, Two independent biochemical pathways for isopentenyl diphosphate and isoprenoid biosynthesis in higher plants, Physiol. Plant. 101 (1997) 643–652.
[67] A. Ren, et al., Molecular characterization and expression analysis of GlHMGS, a gene encoding hydroxymethylglutaryl-CoA synthase from *Ganoderma lucidum* (Ling-zhi) in ganoderic acid biosynthesis pathway, World J. Microbiol. Biotechnol. 29 (3) (2013) 523–531.
[68] J.W. Xu, Y.N. Xu, J.J. Zhong, Production of individual ganoderic acids and expression of biosynthetic genes in liquid static and shaking cultures of *Ganoderma lucidum*, Appl. Microbiol. Biotechnol. 85 (4) (2010) 941–948.
[69] M.S. Shiao, Triterpenoid natural products in the fungus *Ganoderma lucidum*, J. Chin. Chem. Soc. 39 (1992) 669–674.
[70] A. Ren, et al., Profiling and quantifying differential gene transcription provide insights into ganoderic acid biosynthesis in *Ganoderma lucidum* in response to methyl jasmonate, PLoS One 8 (6) (2013) e65027.
[71] C.X. Liang, et al., Enhanced biosynthetic gene expressions and production of ganoderic acids in static liquid culture of *Ganoderma lucidum* under phenobarbital induction, Appl. Microbiol. Biotechnol. 86 (5) (2010) 1367–1374.
[72] B.J. You, et al., A novel approach to enhancing ganoderic acid production by *Ganoderma lucidum* using apoptosis induction, PLoS One 8 (1) (2013) e53616.
[73] W.X. Zhang, Y.J. Tang, J.J. Zhong, Impact of oxygen level in gaseous phase on gene transcription and ganoderic acid biosynthesis in liquid static cultures of *Ganoderma lucidum*, Bioprocess Biosyst. Eng. 33 (6) (2010) 683–690.
[74] H.L. Yang, Ganoderic acid produced from submerged culture of *Ganoderma lucidum* induces cell cycle arrest and cytotoxicity in human hepatoma cell line BEL7402, Biotechnol. Lett. 27 (12) (2005) 835–838.
[75] J. Jiang, et al., Ganoderic acids suppress growth and invasive behavior of breast cancer cells by modulating AP-1 and NF-kappaB signaling, Int. J. Mol. Med. 21 (5) (2008) 577–584.
[76] A. Thyagarajan, et al., Triterpenes from *Ganoderma lucidum* induce autophagy in colon cancer through the inhibition of p38 mitogen-activated kinase (p38 MAPK), Nutr. Cancer 62 (5) (2010) 630–640.
[77] T.S. Wu, L.S. Shi, S.C. Kuo, Cytotoxicity of *Ganoderma lucidum* triterpenes, J. Nat. Prod. 64 (8) (2001) 1121–1122.
[78] R.M. Liu, J.J. Zhong, Ganoderic acid Mf and S induce mitochondria mediated apoptosis in human cervical carcinoma HeLa cells, Phytomedicine 18 (5) (2011) 349–355.
[79] W. Tang, et al., Ganoderic acid T from *Ganoderma lucidum* mycelia induces mitochondria mediated apoptosis in lung cancer cells, Life Sci. 80 (3) (2006) 205–211.
[80] C.H. Li, et al., Ganoderic acid X, a lanostanoid triterpene, inhibits topoisomerases and induces apoptosis of cancer cells, Life Sci. 77 (3) (2005) 252–265.
[81] H.S. Zhu, et al., Effects of extracts from sporoderm-broken spores of *Ganoderma lucidum* on HeLa cells, Cell Biol. Toxicol. 16 (3) (2000) 201–206.

[82] H. Hu, et al., *Ganoderma lucidum* extract induces cell cycle arrest and apoptosis in MCF-7 human breast cancer cell, Int. J. Cancer 102 (2002) 250–253.
[83] Q.X. Yue, et al., Proteomics characterization of the cytotoxicity mechanism of ganoderic acid D and computer-automated estimation of the possible drug target network, Mol. Cell. Proteomics 7 (5) (2008) 949–961.
[84] S.S. Teow, et al., Recent Progress in *Ganoderma lucidum* Research, The Pharmaceutical Society of Korea, Seoul, 1997, pp. 21–39.
[85] C. Hsieh, F.C. Yang, Reusing soy residue for the solidstate fermentation of *Ganoderma lucidum*, Bioresour. Technol. 91 (2004) 105–109.
[86] Y.S. Song, et al., Antiangiogenic and inhibitory activity on inducible nitric oxide production of the mushroom *Ganoderma lucidum*, J. Ethnopharmacol. 90 (2004) 17–20.
[87] J. Jiang, et al., *Ganoderma lucidum* inhibits proliferation and induces apoptosis in human prostate cancer cells PC-3, Int. J. Oncol. 24 (2004) 1093–1100.
[88] K.C. Kim, et al., Enhanced induction of mitochondrial damage and apoptosis in human leukemia HL-60 cells by the *Ganoderma lucidum* and *Duchesnea chrysantha* extracts, Cancer Lett. 246 (2007) 210–217.
[89] D.E. Kleiner, W.G. Stetler-Stevenson, Matrix metalloproteinases and metastasis, Cancer Chemother. Pharmacol. 43 (1999) S41–S51.
[90] A. Scorilas, et al., Overexpression of matrix metalloproteinase-9 in human breast cancer: A potential favourable indicator in node-negative patients, Br. J. Cancer 84 (2001) 1488–1496.
[91] N.H. Chen, J.W. Liu, J.J. Zhong, Ganoderic acid T inhibits tumor invasion in vitro and in vivo through inhibition of MMP expression, Pharmacol. Rep. 62 (1) (2010) 150–163.
[92] C.J. Weng, et al., Inhibitory effects of *Ganoderma lucidum* on tumorigenesis and metastasis of human hepatoma cells in cells and animal models, J. Agric. Food Chem. 57 (11) (2009) 5049–5057.
[93] H. Tilg, et al., Serum levels of cytokines in chronic liver diseases, Gastroenterology 103 (1) (1992) 264–274.
[94] S. Tamir, S.R. Tannenbaum, The role of nitric oxide ($NO^{\bullet}$) in the carcinogenic process, Biochim. Biophys. Acta 1288 (2) (1996) F31–F36.
[95] P. Isomäki, J. Punnonen, Pro- and anti-inflammatory cytokines in rheumatoid arthritis, Ann. Med. 29 (6) (1997) 499–507.
[96] R.K. Coker, G.J. Laurent, Pulmonary fibrosis: Cytokines in the balance, Eur. Respir. J. 11 (6) (1998) 1218–1221.
[97] J. Liu, et al., Ganoderic acid DM: Anti-androgenic osteoclastogenesis inhibitor, Bioorg. Med. Chem. Lett. 19 (8) (2009) 2154–2157.
[98] J.J. Gao, et al., Quantitative determination of bitter principles in specimens of *Ganoderma lucidum* using high-performance liquid chromatography and its application to the evaluation of ganoderma products, Chem. Pharm. Bull. 52 (6) (2004) 688–695.
[99] X.M. Wang, et al., Quantitative determination of six major triterpenoids in *Ganoderma lucidum* and related species by high performance liquid chromatography, J. Pharm. Biomed. Anal. 41 (3) (2006) 838–844.
[100] C. Cheng, et al., Identification of metabolites of ganoderic acid D by ultra-performance liquid chromatography/quadrupole time-of-flight mass spectrometry, Drug Metab. Dispos. 40 (12) (2012) 2307–2314.
[101] S. Teekachunhatean, et al., Pharmacokinetics of ganoderic acids a and f after oral administration of ling zhi preparation in healthy male volunteers, Evid. Based Complement. Altern. Med. 2012 (2012) 780892.

CHAPTER FOUR

Anticancer Effect and Molecular Targets of Saffron Carotenoids

S. Zahra Bathaie*,†,1, Azam Bolhassani‡, Fuyuhiko Tamanoi†

*Department of Clinical Biochemistry, Tarbiat Modares University, Tehran, Iran
†Department of Microbiology, Immunology and Molecular Genetics, Jonsson Comprehensive Cancer Center, Molecular Biology Institute, University of California, Los Angeles, California, USA
‡Department of Hepatitis and AIDS, Pasteur Institute of Iran, Tehran, Iran
[1]Corresponding author: e-mail address: bathai_z@modares.ac.ir; zbatha2000@yahoo.com

Contents

Abstract

Saffron carotenoids, crocin and crocetin, have shown anticancer activity in various animal models of cancer and against different cancerous cell lines. The radical scavenging property and activation of antioxidant defense system are two well-known characteristics of these compounds. However, the results of the studies indicated other mechanisms could also be involved in this function. Insights into various molecular mechanisms of action for crocin and crocetin have been obtained in recent years. The results indicated that despite the structural similarity of crocin and crocetin, their anticancer effects may exert through different mechanisms. Particular interest concerns the ROS-dependent signaling pathways of crocetin. Saffron compounds are safe and may provide inexpensive therapy for treating cancer. They also have protective

The Enzymes, Volume 36
ISSN 1874-6047
http://dx.doi.org/10.1016/B978-0-12-802215-3.00004-5

potential in targeting other disorders including diabetes, Alzheimer's and cardiovascular disease, cognitive deficits, ischemia-induced retinal damage, and many other diseases.

LIST OF THE CELL LINES USED IN THIS REVIEW

22rv1 hormone-refractory prostate cancer cells
A-549 human alveolar basal epithelial cells (non-small cell lung adenocarcinoma)
AGS derived of a biopsy specimen of an untreated human stomach adenocarcinoma
BPH-1 hyperplastic epithelial cells
BxPC-3 pancreatic cancer cell line
C3H10T1/2 multipotential mesenchymal cells derived from C3H mouse embryos
C4-2B osteotropic prostate cancer cells derived from bone metastatic LNCaP cells
Caco-2 adherent human colon cancer cells
CCRF-CEM human T-cell lymphoma
CEM/ADR5000 doxorubicin-resistant subline (colon cancer) cells
CWR22 androgen-dependent prostate cancer cells
DHD/K12-PROb human colon adenocarcinoma cells
DLA Dalton's lymphoma ascite cells
DU145 androgen receptor-negative prostate cancer cells
EAC Ehrlich ascite carcinoma cells
EPN normal epithelial cells
Fibrosarcoma 1456 human primary cells obtained from surgical sample
HCT-116 colorectal cancer cells
HSFS-PI3 human normal fibroblast skin cells
HeLa human epithelioid cervical carcinoma cells
HT-29 human colon adenocarcinoma cells
L929 mouse fibroblast cell line
LAPC-4 androgen-dependent prostate cancer cells
LnCaP androgen-dependent prostate cancer cells
MCF-7 ER^+, PR^+ breast cancer cell line
MCF-7/VCR vincristine-resistant breast cancer cell line
MDA-MB-231 triple negative breast cancer cell line
MIA-PaCa-2 pancreatic cancer cells
MG63 *Homo sapiens* bone osteosarcoma cells
NIH/3T3 mouse fibroblast cells
Ovarian Carcinoma 1998 human primary cells obtained from surgical sample
OS732 osteosarcoma cells
Osteosarcoma 917 human primary cells obtained from surgical sample
P388 leukemia cell line
PC3 androgen receptor-negative prostate cancer cells
RD human rhabdomyosarcoma cells
S-180 sarcoma 180 cell line
SKOV3 ovarian cancer cell line
SW-480 human colorectal cancer cells
T24 human bladder carcinoma cells
TC-1 primary lung epithelial cells of C57BL/6 mice
TCC-56371 human transitional cell carcinoma

VA-13 fetal lung fibroblast, WI-38, cells transformed *in vitro* by SV40 tumor virus
WI-38 normal lung fibroblasts cells

1. INTRODUCTION

Natural products including saffron have long been used to prevent and treat various diseases including cancer and might be the good candidates for the development of anticancer drugs [1,2]. The most important part of saffron is stigma. The extract of saffron stigma consists of various compounds such as carotenoids [3–5]. The origin of wild-type saffron dates back to more than 3000 BC in the regions now known as Greece and Iran. The earliest apparent reference to its cultivation goes back to about 2300 BC in southern Mesopotamia (Sumer), the Accadian empire, on the Euphrates called Azupirano—the name meaning perhaps "Saffron Town" [6,7]. Nowadays, it is cultivated in many countries, but more than 85% of the worldwide saffron production is in Iran [1,2]. The saffron stigmas contain three main metabolites: (a) crocins (a 20-C atoms backbone with various degrees of glucosylation that cause their solubility in aqueous media), (b) picrocrocin responsible for saffron's bitter taste, and (c) safranal responsible for the typical saffron aroma [3–5].

2. ANTICANCER EFFECT OF SAFFRON AND ITS CAROTENOIDS

According to our literature survey, in the modern world, the first report about the antitumor effect of saffron and its active component, crocetin was in 1976 by Gainer *et al.* on skin papillomas and rous sarcoma [8]. In that study, the effectiveness of crocetin to both decrease the number of tumors and delay the onset was reported. This effect was also confirmed in the next studies on skin cancer in hairless mice [9], and topical application of crocetin on 12-*O*-tetradecanoylphorbol-13-acetate (TPA)-induced skin tumors in mice was recommended [10]. The effect of various carotenoids, including four retinoids, namely, retinal, all-*trans* retinoic acid (ATRA), retinyl acetate, retinyl palmitate, and three carotenoid including β-carotene, lycopene, and crocetin, on transplanted C-6 glioma cells on rat was reported in 1989 by Wang *et al.* [11,12]. These results suggested that retinoids or carotenoids are biologically active as antitumor agents against brain tumor cells [12].

The cytotoxic action of saffron ethanolic extract (SEE) on a variety of tumor cells, including: S-180, P388, EAC, and DLA in the *in vitro* were studied. The IC_{50} for all cells are shown in Table 4.1 [13,24]. Crocin, crocetin, and dimethylcrocetin inhibited the growth of HL-60 leukemia cells at a lower dose (Tables 4.2 and 4.3) than ATRA (0.12 μM) [25]. All these compounds at 5 μM induce various degrees of differentiation in the named cells [25]. Oral administration of SEE demonstrated an inhibitory effect on two-stage carcinogenesis of mouse skin papillomas, using dimethylbenz[a]anthracene (DMBA) as an initiator and TPA as a promoter. Studies using purified components indicated that α-crocin was more potent than crocetin gentiobiosyl glucosyl ester and crocetin di-glucosyl ester in inhibiting the tumor promotion [38]. Long-term weekly injection (up to 13 weeks) of crocin (400 mg/kg) to colon adenocarcinoma-induced rats significantly increased the survival of female rats with no major toxic effect [28]. The inhibitory effect of crocetin on the growth inhibition of RD cells was reported [39]. The anticancer effect of saffron and its constituents against HeLa, A-204 and HepG-2 has been reported. Crocins were more potent compounds [40]. In another study, MIA-PaCa-2 cells were injected into the athymic (nude) mice for the development of tumor, and the tumor size and angiogenesis were evaluated with and without crocetin treatment. This series of experiments confirms that crocetin inhibits cellular growth and angiogenesis and stimulates cell death in pancreatic cancer cells [41].

After these observations and confirmation of the anticancer effect of saffron and its carotenoids, the important question was about the mechanism of the anticancer effect. One of the first mechanisms that has been suggested for saffron carotenoids action was the possible provitamin A activity of crocin [24]. However, in the next studies, it has been shown that saffron carotenoids are not provitamin A precursors [25]. In later sections, we will discuss the findings in published data regarding mechanisms involved in saffron biological functions. Fig. 4.1 shows the molecular mechanisms involved in saffron carotenoids anticancer effects.

3. COMPARING THE EFFICACY OF CROCETIN, CROCIN, AND OTHER COMPONENTS

Our study suggests that crocetin in general is more potent than crocin. The effect of crocetin was detected in the concentrations much lower than SAE or crocin. Some of the important examples are our observations about the effectiveness of SAE, crocin, and crocetin in AGS cells and MNNG-induced gastric cancer in rat [36,37,42], as well as in the *N*-nitroso-*N*-methylurea (NMU)-induced breast cancer [43]. Our results indicated that crocetin (IC_{50} 100 μM) was more effective than crocin

Table 4.1 Cytotoxic Effects of Saffron Extract on Tumor Cell Lines

Compounds	Cells	IC_{50}	LD_{50}	References
Saffron ethanolic extract (SEE)	S-180	28 μg/ml		[13]
	P388	30 μg/ml		
	EAC	17 μg/ml		
	DLA	9 μg/ml		
	HeLa		2.3 mM	[14]
		800 μg/ml		[15]
	HepG2	950 μg/ml		
	MCF-7	400 ± 18.5 μg/ml		[16]
	A549	650 μg/ml		[17]
	LAPC-4	0.8 ± 0.2 mg/ml		[18]
	CWR22	1.3 ± 0.7 mg/ml		
	LnCaP	1.4 ± 0.6 mg/ml		
	22rv1	2.3 ± 0.5 mg/ml		
	C4-2B	2.7 ± 0.3 mg/ml		
	DU145	5.3 ± 1.2 mg/ml		
	PC3	7.4 ± 0.5 mg/ml		
	EPN	NE mg/ml		
	BPH-1	NE mg/ml		
Saffron extract	HeLa; A549; WI-38VA	100–250 μg/ml		[19,20]
	A-204; HEPG-2; SW-480	150–200 μg/ml		[21]
Saffron aqueous extract (SAE)	TC-1	4 mg/ml		[22]
	TCC-5637l	~300 μg/ml		[23]
	L929	> 600 μg/ml		[23]
	LAPC-4	4.2 ± 0.7 mg/ml		[18]
	CWR22	4.5 ± 0.4 mg/ml		
	LnCaP	4.0 ± 0.8 mg/ml		

Continued

Table 4.1 Cytotoxic Effects of Saffron Extract on Tumor Cell Lines—cont'd

Compounds	Cells	IC_{50}	LD_{50}	References
	22rv1	6.5 ± 1.2 mg/ml		
	C4-2B	5.8 ± 0.5 mg/ml		
	DU145	>0.8 mg/ml		
	PC3	>0.8 mg/ml		
	EPN	NE mg/ml		
	BPH-1	NE mg/ml		

Most of the data presented in the table are after 48 h treatment of the cells.
IC_{50}: The drug concentration for 50% cytotoxicity.
LD_{50}: The doses inducing 50% cell growth inhibition.

(IC_{50} 2760 μM) to induce apoptosis in AGS cells. Similar results were also obtained in animal models of both gastric and breast cancers. Treatment of the three types of cancer cells: HeLa, A-549, and SKOV3 with crocetin (120–240 μM) for 48 h significantly inhibited their proliferation in a concentration- and time-dependent manner [44]. When crocetin used with vincristine in the mentioned cancer cells, it (60 μM) significantly enhanced the cytotoxicity induced by vincristine (1 μM) [44]. This synergistic effect was also detected in the vincristine-resistant breast cancer cell line MCF-7/VCR. Therefore, crocetin is a potential anticancer agent, which may be used as a chemotherapeutic drug or as a chemosensitizer for vincristine [44]. All of the mentioned studies indicated that the saffron components had no effect on normal cells, like HSFS-PI3, or in normal animals [36, 37, 42, 43].

In other studies, however, crocetin was ineffective, in contrast to crocin on the growth inhibition of HeLa cells [14]. Although crocetin inhibited the polynucleotide biosynthesis, but it had no effect on colony formation in HeLa, A-549, and VA-13 cells after 3 h incubation [45]. In a study, to examine and compare the anticancer properties of saffron and its constituents against HeLa, A-204, and HepG-2, crocins announced as the more potent compound [40]. Incubation of MCF-7 and MDA-MB-231 breast cancer cells for 48 h with different concentrations of SE showed a dose-dependent inhibitory effect on cell proliferation measured by the MTT assay. Studies on the effect of α-crocin and crocetin showed that the antiproliferative effect is attributed to the constituent crocins irrespective of the degree of glycosylation [46]. The apparent difference between the above studies may be due to differences in experimental design and time course of the experiments.

Table 4.2 Cytotoxic Effects of Crocin on Tumor Cell Lines

Compound	Cells	IC_{50}	LD_{50}	References
Crocin	HL-60	2 μM		[25]
	HeLa		3 mM	[14]
		1.603 mM		[26]
		3.58 mM		[27]
	HT-29		0.4 mM	[28]
	DHD/K12-PROb		1.0 mM	[28]
	Caco-2	154.00 ± 8.68 μM		[29]
	CEM/ADR5000	206.30 ± 10.14 μM		
	CCRF-CEM	70.30 ± 4.85 μM		
	HepG2	3 mg/ml		[30]
		2.87 mM		[27]
	LAPC-4	0.26 ± .03 mM		[18]
	CWR22	0.34 ± 0.7 mM		
	LnCaP	0.55 ± 0.11 mM		
	22rv1	0.44 ± 0.01 mM		
	C4-2B	0.48 ± 0.07 mM		
	DU145	0.84 ± 0.13 mM		
	PC3	0.95 ± 0.23 mM		
	EPN	3.95 ± 1.23 mM		
	BPH-1	2.74 ± 1.82 mM		
	DLA	13 μg/ml		[31]
	EAC	19 μg/ml		
	L1210 Leukemia	11 μg/ml		
	P3888 Leukemia	39 μg/ml		
	Sarcoma S-180	19 μg/ml		
	K562	3 mM		[32]
	SW-480	1.0 mM		[33]
	HT-29	1 < X < 3 mM		

Continued

Table 4.2 Cytotoxic Effects of Crocin on Tumor Cell Lines—cont'd

Compound	Cells	IC_{50}	LD_{50}	References
	HCT-116	1<X<3 mM		
		1.99 mM		[27]
	C26 colon carcinoma	0.73 mM		[34]
	BxPC-3	10 μg/ml		[35]
	TC-1	1.5 mM		[22]
	AGS	2.76 mM		[36]
	A549	5.48 mM		[27]
	SKOv3	3.35 mM		

Most of the data presented in the table are after 48 h treatment of the cells.
IC_{50}: The drug concentration for 50% cytotoxicity.
LD_{50}: The doses inducing 50% cell growth inhibition.

Pharmacokinetic data indicated that crocetin absorbed into the blood through the gastrointestinal (GI) tract of rat and mice [47,48]. However, pharmacokinetic study of crocin in rat indicated that orally administered crocin did not absorb through GI tract and a large part of administered crocin is excreted into the feces. The intestinal tract of both mice and rat served as an important site for crocin hydrolysis and it was absorbed to the plasma as crocetin [48,49]. In addition to crocetin, its mono- and di-glucuronide conjugates was also detected in mice plasma [48]. Although, the similar study has not been done in the human, but crocetin was determined in the human plasma after saffron treatment [50], which is an affirmative reason for crocetin absorption through GI. Therefore, the ineffectiveness or low therapeutic effect of orally administered crocin in some human studies may be related to this mechanism of crocin absorption [51]. Finally, combination of saffron extract (SE) with selenite increased the level of inhibition of the colony formation and nucleic acid synthesis in comparison with cells that were treated with only one of these agents [52].

4. LIPOSOME FORMULATION OF SAFFRON COMPOUNDS

To overcome the problem of poor absorption, in some studies using saffron or crocin, use of liposome was considered. Liposome encapsulation

Table 4.3 Cytotoxic Effects of Crocetin and Dimethylcrocetin on Tumor Cell Lines

Compounds	Cells	IC_{50}	LD_{50}	References
Crocetin	HL-60	2 μM		[25]
	HeLa	No cytotoxic (after 3 h)		[14]
		0.22 mM		[27]
	HepG2	0.61 mM		
	HCT-116	0.16 mM		
	A549	0.41 mM		
	SKOV3	0.19 mM		
	K562	2 μM		[32]
	AGS	100.00±3.14 μM		[37]
Dimethylcrocetin	HL-60	0.8 μM		[25]
	K562	0.8 mM		[32]
	DLA	9 μg/ml		[31]
	EAC	17 μg/ml		
	L1210	9 μg/ml		
	P388	30 μg/ml		
	Sarcoma-180	12 μg/ml		
	Osteosarcoma 917	14 μg/ml		
	Fibrosarcoma 1456	7 μg/ml		
	Ovarian Carcinoma 1998	8 μg/ml		

Most of the data presented in the table are after 48 h treatment of the cells.
IC_{50}: The drug concentration for 50% cytotoxicity.
LD_{50}: The doses inducing 50% cell growth inhibition.

of saffron increased its antitumor activity against S-180 and EAC solid tumor [53]. Furthermore, crocin and its liposomes induced cell death in HeLa and MCF-7 cells, in which liposomal encapsulation improved cytotoxic effects. Crocin liposomal forms (IC_{50} values after 48 h: 0.61, 0.64, and 1.2 mM) showed enhanced cytotoxic effect compared with the crocin (IC_{50} after 48 h: 1.603 mM) in HeLa cells [26].

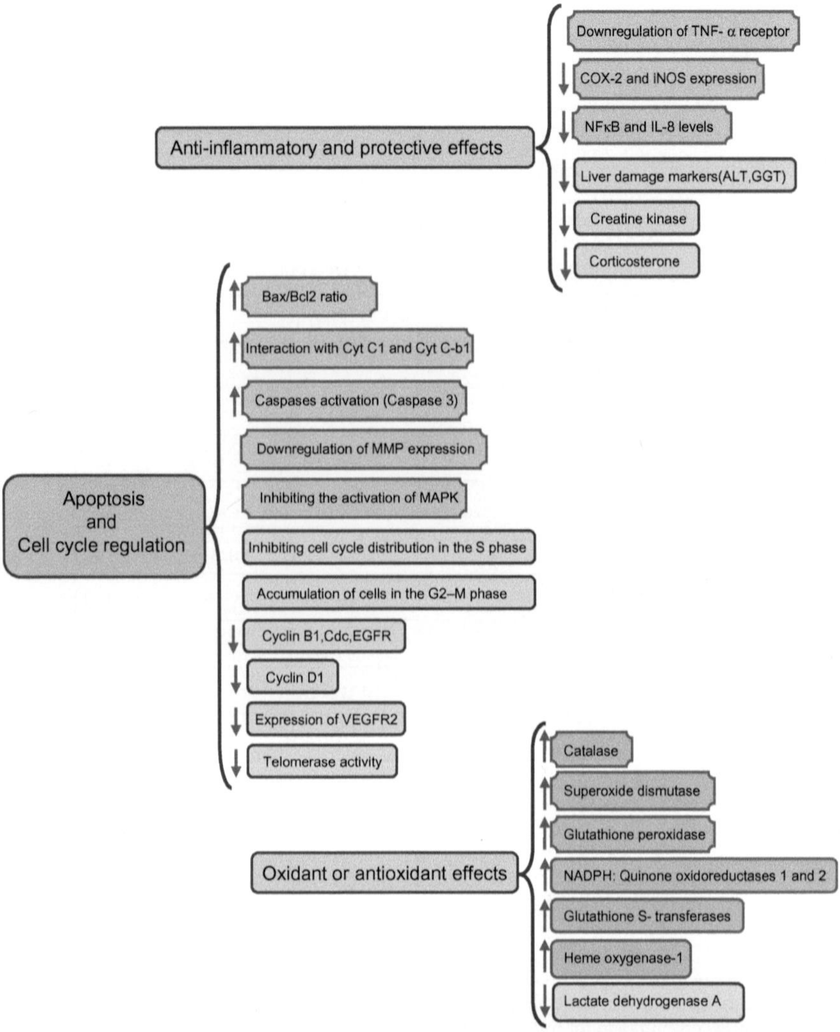

Figure 4.1 Some of the mechanisms involved in antitumor effects of saffron carotenoids. The effect of crocin and crocetin on ROS production, p53 and p21 expression is completely different, which indicated the difference in the mechanism of action. The issues should be more investigated in future. (See the color plate.)

5. EFFECT OF CROCETIN AND CROCIN ON MACROMOLECULE SYNTHESIS AND STRUCTURE

5.1. Effect on DNA, RNA, and Protein Synthesis

The effect of SE on macromolecular synthesis in three human cell lines: A-549, WI-38, and VA-13 cells [19] and HeLa cells [20] were investigated.

The results showed that the malignant cells were more sensitive than the normal cells to the inhibitory effects of SE on both DNA and RNA synthesis. There was no effect on protein synthesis in any of the cells, even at higher SE concentration [19,20]. The cellular DNA and RNA content was also significantly downregulated by 0.4 mM crocin in human tongue squamous cell carcinoma cell line, Tca8113 [54]. While in another study, incubation of HeLa, A-549, and VA-13 cells with crocetin for 3 h caused a dose-dependent inhibition of nucleic acid (DNA and RNA) and protein synthesis, but not on colony formation [45]. Thymidine incorporation assay indicated that DNA synthesis was inhibited in the P388 leukemia cells by SEE at concentration 4.2 μg/ml, which is lower than IC_{50}; it means that the mechanism of action of SEE is at the level of DNA synthesis [13]. Crocetin had a dose-dependent inhibitory effect on DNA, RNA, and protein synthesis in isolated nuclei and suppressed the activity of purified RNA polymerase II [45].

The report of SE effect on DNA synthesis was followed by a report indicating direct interactions of SE and all of saffron carotenoids with high molecular weight DNA *in vitro*. There were no preferences for A·T- or G·C-rich oligonucleotides [55]. The interaction of tRNA with saffron carotenoids was also shown [56]. We also showed the *in vitro* interaction of saffron carotenoids with DNA Tetraplexes (G-quadruplex and i-motif) [57]. Crocetin treatment decreased the benzo(α)pyrene (BαP) DNA adduct formation in C3H10T1/2 cells [58]. In addition, protective effect of saffron carotenoids against DNA damage induced by chemicals were shown by comet assay [59,60]. All of these results confirm the direct interaction of nucleotides with crocetin in the cell.

5.2. Protein Binding

Crocetin is not soluble in the aqueous solution at the physiologic pH. Thus, binding of crocetin to plasma albumin was studied for the first time at 1979 [61] as a mechanism for its transport in the body fluids; and then continued to obtain more information about the nature of this interaction [62,63]. Interaction of other saffron constituents was also shown with albumin [64]. In an *in vitro* experiment, we also showed the interaction of saffron carotenoids with histone H1, induction of conformational changes on it, and a decrease in the DNA–histone H1 complex formation [65]. Involvement of this interaction, as a possible mechanism for the observed changes in the expression of various genes, should be investigated *in vivo*. Crocin increased tubulin polymerization and microtubule nucleation rate in a concentration dependant manner [66]. The interaction of crocin with the amyloid-beta peptides 1–40 (Aβ-40) [67] and Aβ-42 [68], as well as its preventive effect on amyloid fibril formation have been shown by us. Various methods including fluorescence spectroscopy, CD spectroscopy, and electron microscopy were used. Crocin interacts with Aβ peptides and prevents amyloid formation. In addition, crocin could disrupt the amyloid fibrils [68]. This means that it has the potential to be an important therapeutic drug against Alzheimer's disease. Crocetin also inhibited Aβ fibril formation and destabilized preformed Aβ fibrils. Moreover, crocetin caused stabilization of Aβ oligomers and prevented their conversion into Aβ fibrils [69]. These results indicated that after entering cell, crocin and crocetin not only could bind to DNA and RNA but also interact with various proteins and can modulate their functions. Therefore, saffron carotenoids have the potential to affect several cellular processes.

6. EFFECTS ON CELL CYCLE, APOPTOSIS, AND SIGNALING PATHWAYS

A cDNA microarray study of human bladder carcinoma T24 cell indicated that 836 genes were upregulated or downregulated after crocin treatment, by more than two times [70]. The upregulation of p21(WAF1) and downregulation of cyclin D1 was shown in the levels of both mRNA and protein by crocin in T24 cells [70]. Our results also showed the downregulation of cyclin D1 by crocin in the breast tumor induced by NMU injection in female rat. Crocin also downregulated both p21 Cip1 and p53 expressions in this tumor tissue [43]. Crocin exerts both *in vitro* and *in vivo* (BALB/c xenograft tumor) anticancer effect on TCCB T24 cell line. These cellular effects may alter cell cycle, and induce tumor cell apoptosis by downregulating the expression of B cell lymphoma/leukemia-2 (Bcl-2), survivin, cyclin D1 and upregulating the expression of Bax (decrease Bcl-2/Bax ratio) [71].

The effect of both SEE and crocin on a series of prostate cancer cell lines was investigated [18]. The results demonstrated that crocin induced G0/G1 and G2/M phase cell cycle arrest by downregulating the expression of cyclin D1 in a dose-dependent manner, leading to a decrease in the formation of cyclin D1/CDK-4 holoenzyme complex. Moreover, crocin upregulated CDK inhibitors, such as p21 and p27 [18]. Western blot analysis revealed that the expression of Bcl-2 was strikingly downregulated, whereas Bax was upregulated [18]. We also obtained similar results on the decrease in

Bcl-2/Bax ratio and induction of caspases activity by crocin in AGS cells [36]. SE effect on MCF-7 cells is caspases dependent and accompanied with an increase in Bax protein expression [16].

Analysis of caspases activity indicated a caspase-dependent pathway with involvement of caspase-9 activation, suggesting an intrinsic pathway in the apoptosis induced by crocin in a series of prostate cancer cell lines [18]. However, combination of crocin and cisplatin showed a strong killing effect on osteosarcoma cells and suppresses the ability of invasion of MG63 and OS732 cells, which might be related to upregulate the expression of caspase-3 and caspase-8 [72].

The data presented for different prostatic cancer cells indicated that hormone-sensitive LAPC-4, CWR22, and LnCaP cells are more sensitive (prone) to SE and crocin, when compared to hormone-insensitive 22rv1, C4-2B cells. The IC_{50} between 0.26 and 0.95 mM for cancerous cells and near four for normal cells [18] are presented in Tables 4.1 and 4.2.

Crocin treatment decreased the activity of telomerase in HepG2, which is probably caused by downregulation of the expression of the catalytic subunit of the enzyme. Because the relative expression level of the catalytic subunit of telomerase (hTERT) gene showed a 60% decrease as compared to untreated control cells [30]. Another mechanism involved in the decreased telomerase activity is the direct interaction between crocin and telomeric quadruplex sequences, which was shown in an *in vitro* experiment [57]. The fluorescence resonance energy transfer analysis showed a slight change in the T_m of telomeric oligonucleotides in the presence of crocin [30].

Crocetin has been introduced as a potent inhibitor of tumor promotion induced by TPA in mouse skin [73]. It suppressed the TPA (100 ng/ml)-induced c-Jun and c-Fos gene expression, similar to that observed in the NIH/3T3 (mouse fibroblast) cells pretreated with crocetin [73]. Suppression of the TPA-induced c-Jun, c-Fos, and c-Myc expression by crocetin in the mouse epidermis was also reported [74].

When NIH/3T3 cells were treated with TPA alone, protein kinase C (PKC) translocated from the cytosolic fraction to the particulate fraction. Pretreatment with crocetin inhibited the TPA-induced PKC activity in the particulate fraction, but did not affect the level of PKC protein. Crocetin also reduced the level of TPA-stimulated phosphorylation of cellular proteins [73]. Similar mechanism was also reported elsewhere. These data indicated the role of crocetin as an inhibitor of angiotensin (Ang) II-induced membrane translocation of PKCα, and the inhibitory activity of crocetin on PKC activity in membrane fraction. In addition, AngII-induced mRNA expressions of c-Fos, c-Jun, and c-Myc were also suppressed by crocetin. The effects of crocetin on PKCα translocation and PKC activity, and the subsequent suppression of proto-oncogene expressions might mediate its inhibitory effect on vascular smooth muscle cells (VSMCs) proliferation [75]. In addition, crocetin markedly suppressed AngII-induced activation of extracellular signal-regulated kinase 1/2 (ERK1/2) and its downstream effector c-Fos expression, which is a prerequisite for cell cycle progression. These results might be related, at least partly, to the antioxidant property of crocetin [76]. A Ca^{2+}-dependent pathway may be involved in the inhibitory effect by crocetin of AngII-induced ERK1/2 activation [77]. Crocetin also markedly inhibited the expression of cyclin D1, but not cyclin E. In addition, crocetin increased the level of cyclin-dependent kinase inhibitor (CDKI) p27(kip1), but not CDKI p27(waf1/cip1) in VSMCs [78]. Treatment of human colon cancer SW480 cells with crocetin was also induced cell cycle arrest through p53-independent mechanisms accompanied by P21 induction, in a time- and dose-dependent manner. Crocetin (0.8 mM) caused cytotoxicity in the SW480 cells by enhancing apoptosis and decreasing DNA repair capacity [79]. Crocetin (240 μM) significantly induced cell cycle arrest through p53-dependent and -independent mechanisms accompanied with p21(WAF1/Cip1) induction in HeLa, A549, and SKOV3 cells [44]. We have also recently showed that crocetin suppresses tumor growth in NMU-induced breast cancer through the downregulation of cyclin D1 and upregulation of $p21^{Cip1}$ in a p53-dependent manner [43].

Cancer growth, invasiveness and metastasis have been associated with overexpression of metalloproteinases (MMPs). The effect of crocetin and ATRA on the expression of MMPs and their tissue inhibitors (TIMPs) in the highly invasive epithelial breast cancer cells, MDA-MB-231 indicated that crocetin (1 and 10 μM) and ATRA (10 μM) reduced expression of MT-MMPs and increased TIMP-3 mRNA level, although they do not affect cell proliferation at these concentrations. Only pro-MMP-9 (matrix metalloproteinase-9) was recorded in zymographs [80,81].

Crocetin treatment inhibited tumor size induced due to the injection of MIA-PaCa-2 cells into the athymic nude mice, and histochemical studies showed the inhibition of both VEGF and EGFR. Expression of Bax was increased with concomitant decrease of Bcl-2 (i.e., Bcl-2/Bax ratio decreased) in the tumor developed in nude mice [41]. Both the *in vitro* pancreatic cancer cells (MIA-PaCa-2) and *in vivo* athymic nude mice tumor, apoptosis was significantly stimulated as indicated by Bax/Bcl-2 ratio. Cdc-2, Cdc-25C, Cyclin-B1, and epidermal growth factor receptor were altered significantly by crocetin [41]. Crocetin inhibitory effect on AGS cells was also indicated a decrease in Bcl-2/Bax ratio, and an increase in the caspases activity [37].

Crocetin inhibited c-Jun NH_2-terminal kinase (JNK) and inhibitor κB-kinaseβ activation via suppression of PKCτ phosphorylation, attenuating insulin insensitivity induced by palmitate in 3T3-L1 adipocytes [82]. Crocetin inhibited the rat liver cAMP-dependent protein kinase (PKA) with the IC_{50} of 3.0 μM [83].

Comparison between the effect of SE, crocin, and crocetin on MCF-7 and MDA-MB-231 breast cancer cells showed the antiproliferative effect was not related to the presence of estrogen receptors [46]. In addition, crocetin exerts the anticancer effect at lower concentrations than SE or crocin [27,43]. Effect of SAE, crocin, and crocetin on the rat ovary and brain enkephalin-degrading aminopeptidase activity has been shown in NMU-induced breast cancer [84].

The effect of saffron in HCT-116 colorectal cancer cells with different p53 status; two p53 isogenic HCT116 cell lines (HCT wild type and HCT p53 −/−) has been reported. Saffron induced DNA-damage and apoptosis in both cell lines. In addition, the saffron-induced LC3-II protein level was a remarkable indication of the accumulation of autophagosomes, a response to the cellular stress condition of drug treatment. However, autophagy has delayed the induction of apoptosis in HCT116 p53 −/− cells [85].

Saffron strongly enhanced glucose uptake and the phosphorylation of AMP-activated protein kinase (AMPK)/acetyl-CoA carboxylase (ACC) and mitogen-activated protein kinases (MAPKs), but not PI 3-kinase (phosphatidylinositol 3-kinase)/Akt in C2C12 skeletal muscle cell [86]. Although C2C12 is not a cancerous cell line, but this mechanism should be investigated in tumor cells in future.

The effect of crocin and some other carotenoids for their possible ability to reverse multidrug resistance in resistant Caco-2 and CEM/ADR5000 cells in comparison with the sensitive parent cell line CCRF-CEM was investigated. Crocin and some other carotenoids did not cause substantial cell growth inhibition in resistant cells [29].

7. ROLE OF SAFFRON COMPONENTS ON CHEMOPREVENTION

The chemopreventive effect of crocins [87] and crocetin [88] against aflatoxin B1(AFB1) and dimethylnitrosamine damages in rat liver was shown. Crocetin decreased AFB1-DNA adduct formation in C3H10T1/2 fibroblast cells [89]. Topical application of SE inhibited both stages of initiation and promotion in DMBA/croton oil-induced skin cancer in mice. [90–92]. Oral administration of SE (100 mg/kg body wt) 30 days after subcutaneous administration of 20-methylcholanthrene (MCA) (745 nM × 2 days) restricted tumor incidence up to 10% compared with MCA-treated controls (100%) [90]. Preapplication of crocetin afforded significant protection against TPA-induced hyperplasia in the ear skin of mice [10]. Different doses of saffron protect Swiss albino mice against genotoxins-induced oxidative stress (toxins like cyclophosphamide) by increasing both enzymatic and nonenzymatic liver antioxidant systems [93]. Saffron also protected mice against chromosomal damage induced by cisplatin, mitomycin-C, and urethane [59,94]. Crocetin also prevented the BαP-induced lung cancer in both pre- and postinitiated states [95] in Swiss albino mice by inhibiting glycoprotein and polyamine biosynthesis [96].

One study provided evidence that saffron exerts a significant chemopreventive effect against liver cancer through inhibition of cell proliferation and induction of apoptosis. It also showed that saffron protected rat liver via modulating oxidative damage and suppressing inflammatory response [97]. Administration of saffron (75, 150, and 300 mg/kg/day) 2 weeks prior to the diethylnitrosamine (DEN) injection up to 22 weeks, significantly reduced the DEN-induced increase in the number and the incidence of hepatic dyschromatic nodules. Saffron also decreased the number and the area of placental glutathione *S*-transferase (GST) positive foci in livers of DEN-treated rats [97]. Furthermore, saffron restored superoxide dismutase (SOD), catalase (CAT), and GST levels and diminished myeloperoxidase (MP) activity, malondialdehyde (MDA) and protein carbonyl formation in rat liver [97]. The immunohistochemical staining of rat liver showed that saffron inhibited the DEN-mediated elevations in numbers of cells positive for Ki-67, cyclooxygenase 2, inducible nitric oxide synthase, nuclear factor-kappa B (NFκB) p-65, and phosphorylated tumor necrosis factor receptor. Saffron also blocked the depletion in the number of cells positive for terminal deoxynucleotidyl transferase mediated deoxyuridine triphosphate nick-end labeling (TUNEL) and M30 Cyto-Death in liver tissues of DEN-treated rats. *In vitro* experiments using HepG2 cells also confirmed the above mentioned findings and showed inhibition of NFκB activation, increased cleavage of caspase-3, as well as DNA damage and cell cycle arrest upon saffron treatment [97].

In contrast to the above mentioned reports, a few reports also indicated the ineffectiveness of crocin genotoxicity induced by diazinon [98] or crocetin against HeLa cells [14].

8. MOLECULAR MECHANISMS INVOLVED IN THE PROTECTIVE EFFECT OF SAFFRON COMPONENTS AGAINST VARIOUS DAMAGES IN DIFFERENT TISSUES

The effect of SAE and α-crocin on methyl methanesulfonate-induced DNA damage of different tissues of mice was investigated using the comet assay. The suppression of DNA damage by SAE and crocin was a dose-dependent phenomenon in liver, lung, kidney, and spleen. The data indicated a genoprotective mechanism of SAE and crocin [60].

Treatment with SE or crocin blocked the ability of chronic stress to impair spatial learning and memory retention. The significantly higher levels of lipid peroxidation products, and antioxidant enzymes including: glutathione peroxidase (GSH-px), glutathione reductase (GHR), and SOD and significantly lower total antioxidant reactivity capacity were detected in stressed animals that received SE or crocin. Crocin also significantly decreased plasma level of corticosterone, as measured after the end of stress. These observations indicated that saffron and its active constituent crocin can prevent the impairment of learning and memory as well as the oxidative stress damage to the hippocampus induced by chronic stress [99].

The protective effect of crocin against cisplatin-induced acute renal failure and relative oxidative stress has been reported [100,101]. Crocin treatment resulted in a significant and dose-dependent reduction in MDA concentration and a significant elevation in total thiol and GSH-px concentrations, as compared with cisplatin-treated rats [100].

H_2O_2 decreases the Bcl-2/Bax ratio and increases intracellular Ca^{2+} concentration in bovine aortic endothelial cell. Crocin treatment increases this ratio and inhibited intracellular Ca^{2+} elevation. Thus, crocin inhibits H_2O_2-induced apoptosis in these cells [102]. Crocin showed the similar effect in PC12 cells. Also, crocin inhibited ROS generation in cells exposed to ACR. Thus, pretreatment with crocin protected cells from ACR-induced apoptosis partly by inhibition of intracellular ROS production [103].

The effect of crocin on ischemia/reperfusion (I/R) injury in mice cerebral microvessels was studied. Transient global cerebral I/R, significantly promoted the generation of nitric oxide (NO) and MDA in cortical microvascular homogenates, as well as markedly reduced the activities of SOD and GSH-px and promoted the activity of nitric oxide synthase (NOS). Reperfusion led to serous edema with substantial microvilli loss, vacuolation, membrane damage, and mitochondrial injuries in cortical microvascular endothelial cells (CMEC). Furthermore, enhanced phosphorylation of ERK1/2 and decreased expression of MMP-9 were detected in CMEC after I/R. Reperfusion for 24 h also induced membrane (functional) G protein-coupled receptor kinase 2 (GRK2) expression, while it reduced cytosol GRK2 expression. Pretreatment with crocin markedly inhibited oxidizing reactions and modulated the ultrastructure of CMEC in mice. Furthermore, crocin inhibited GRK2 translocation from the cytosol to the membrane and reduced ERK1/2 phosphorylation and MMP-9 expression in cortical microvessels. Therefore, crocin protects the brain against excessive oxidative stress [104].

Endoplasmic reticulum (ER) stress is a homeostatic mechanism, which is used by cells to adapt to intercellular and intracellular changes. Moreover, ER stress is closely linked to inflammatory pathways [105]. In addition, ER stress is an integral component of neuroinflammation and contributes to the development of neurological diseases. In autopsied brain specimens from multiple sclerosis (MS) and non-MS patients, XBP-1 spliced variant (XBP-1/s) was increased in MS brains and was correlated with the expression of the human endogenous retrovirus-W envelope transcript, which encodes the glycoprotein, Syncytin-1. In primary human fetal astrocytes transfected with a Syncytin-1-expressing plasmid, XBP-1/s, BiP, and NOS2 were induced, which was suppressed by crocin treatment. Crocin also protected oligodendrocytes exposed to cytotoxic supernatants derived from Syncytin-1-expressing astrocytes and NO-mediated oligodendrocytotoxicity. During experimental autoimmune encephalomyelitis (EAE), the transcript levels of the ER stress genes XBP-1/s, BiP, PERK, and CHOP were increased in diseased spinal cords compared with healthy littermates, although CHOP expression was not involved in the EAE disease phenotype [105]. Daily treatment with crocin starting on day 7 post-EAE induction suppressed ER stress and inflammatory gene expression in spinal cords, which was accompanied by preserved myelination and axonal density, together with reduced T-cell infiltration and macrophage activation. EAE-associated neurobehavioral deficits were also ameliorated by crocin treatment. These findings underscored the convergent roles of pathogenic ER stress and immune pathways in neuroinflammatory disease and point to potential therapeutic applications for crocin [105]. These results are consistent with our findings about the improvement effect of crocin on locomotor function and mechanical behavior of the rats with contused spinal cord injury. A decrease in the calcitonin gene-related peptide expression was also observed in these rats [106].

Crocin and crocetin were shown to be effective in the inhibition of lipopolysaccharide (LPS)-induced NO release from cultured rat brain microglial cells. These compounds reduced the LPS-stimulated productions of tumor necrosis factor-alpha (TNFα), interleukin (IL)-1 beta, and intracellular ROS. The compounds also effectively reduced LPS-elicited NFκB activation. In addition, crocin reduced NO release from microglia stimulated with interferon-gamma (IFN-γ) and amyloid-beta. In organotypic hippocampal slice cultures, both crocin and crocetin blocked the effect of LPS on hippocampal cell death. These results suggest that crocin and crocetin provide neuroprotection by reducing the production of various neurotoxic molecules from activated microglia [107].

Pretreatment with crocetin reduced the LPS-induced lung edema and histological changes, increased LPS-impaired SOD activity, and decreased lung myeloperoxidase (MP) activity in mice [108]. In addition, crocetin significantly attenuated LPS-induced mRNA and the protein expressions of interleukin-6 (IL-6), macrophage chemoattractant protein-1, and TNFα. Crocetin at different dosages, also reduced phospho-I kappa B expression and NFκB activity in LPS-induced lung tissue alteration [108].

Crocetin prevented high glucose-induced apoptosis in human umbilical vein endothelial cells (HUVECs), which correlates with the increase of activation of p-Akt, following the upregulation of eNOS and NO production. Pretreatment with phosphatidylinositol 3′ kinase (PI3K) inhibitor LY294002 or eNOS inhibitor NG-nitro-arginine methyl ester (L-NAME) inhibited crocetin effect on p-Akt or eNOS, respectively. Thus, crocetin inhibits high glucose-induced apoptosis, at least partly, via PI3K/Akt/eNOS pathway in HUVECs [109]. Similar protective effect of SE and crocin was shown in PC12 cells [110].

The favorable impacts of crocetin on the inhibition of platelet activity and thrombosis formation may be related to the inhibition of Ca^{2+} elevation in stimulated platelets [111]. The inhibitory effect of crocin was also shown on platelet aggregation and oxidative stress-induced apoptosis [112].

The inhibitory effect of crocetin on cardiac hypertrophy is mediated by blocking the ROS-dependent mitogen-activated protein kinase (MAPK)/extracellular signal-regulated kinase-1/2 (MEK/ERK1/2) pathway and GATA-binding protein 4 activation. Furthermore, crocetin inhibited inflammation by blocking NFκB signaling and attenuated fibrosis and collagen synthesis by abrogating MEK–ERK1/2 signaling [113].

The protective role of crocetin following cerebral contusion and its effects on the enhancement of angiogenesis in rats was proposed [114]. The results revealed that crocetin treatment led to an inhibition of neuronal apoptosis 72 h following treatment; this finding was confirmed by Bcl-2 protein expression. Expression levels of VEGFR-2 and serum response factor were higher in the group treated with crocetin [114].

The protective effect of crocetin against angiogenesis induced by VEGF has also been reported. Crocetin significantly suppressed VEGF-induced tube formation by HUVECs and migration of human retinal microvascular endothelial cells [115]. It also significantly inhibited phosphorylation of p38 and protected VE-cadherin expression. These findings indicate that crocetin suppresses the VEGF-induced angiogenesis by inhibiting migration and that the inhibition of phosphorylated-p38 and protection of VE-cadherin expression may be involved in its underlying mechanism of action [115].

Crocetin administration to experimental animals during resuscitation post hemorrhage increased survival, at least in part by protecting the liver from activation of apoptotic cell death [116].

Crocetin significantly inhibited photoreceptor degeneration and retinal dysfunction and halved the expression of TUNEL-positive cells. Thus, crocetin has protective effects against retinal damage *in vitro* and *in vivo*. The suggested mechanism is suppressing the increased caspase-3 and -9 activities after retinal damage [117].

The effects of crocetin on I/R-induced retinal damage were examined by histological, electrophysiological, and antiapoptotic analyzes. Data for antiapoptotic analysis was obtained by TUNEL staining. Using immunohistochemistry and immunoblotting, the protective mechanism of the crocetin through the expression of 8-hydroxy-2-deoxyguanosine (8-OHdG; a marker of oxidative stress) and on phosphorylations of MAPK, ERK, JNK, p38, NFκB, and c-Jun [118]. The histological analysis revealed that I/R decreased the cell number in the ganglion cell layer (GCL) and the thickness of inner nuclear layer (INL), and that crocetin inhibited GCL and INL. ERG measurements revealed that crocetin prevented the I/R-induced reductions in a- and b-wave amplitudes. In addition, crocetin decreased the numbers of TUNEL-positive cells and 8-OHdG-positive cells, and the phosphorylation levels of p38, JNK, NFκB, and c-Jun present in the retina after I/R [118].

9. ANTIOXIDANT AND ANTI-INFLAMMATORY EFFECTS OF SAFFRON

As many other natural products, saffron is also known for its antioxidant potency from a long time ago. The inhibition of the papillomas in the mouse appeared to be at least partly due on modulatory effects of saffron on some phase II detoxifying enzymes like GST and GPx, as well as CAT and SOD [91]. Standard histological examination of skin demonstrated a beneficial action of saffron in mice, where saffron treatments were given both before and after the induction of skin carcinogenesis. Saffron ingestion inhibited the formation of skin papillomas in animals and simultaneously reduced their size. Thus, saffron inhibits DMBA-induced skin carcinoma in mice when treated early. This may be due, at least in part, to the induction of cellular defense systems [91,92].

The intracellular GSH and related enzymes, i.e., GHR and GST, of S-180 tumor cells were significantly elevated when incubated with saffron [53]. An increase in the activity of GST due to the pretreatment of C3H10T1/2 cells with crocetin against genotoxicity and neoplastic transformation induced by BαP was reported [58].

Crocin showed a high radical scavenging activity [119]. Crocetin also changed the level of antioxidant enzyme in the BαP-induced lung cancer in mice [95].

ROS were generated both enzymatically and nonenzymatically using the xanthine/xanthine oxidase (X/XO) system and paraquat system, respectively in rat primary hepatocytes. The addition of crocetin decreased genotoxicity in both systems. The data showed that crocetin also inhibited the formation of superoxide anion in the X/XO system and bleached the free radical 1,1-diphenyl-2-picrylhydrazyl. The protective action of crocetin operated via quenching of the superoxide anion and/or free radical [120].

Crocetin significantly decreased AngII-induced intracellular reactive oxygen species and increased the activity of SOD in BASMCs [76].

The effects of macerated extracts of *Crocus sativus* (Family Iridaceae) (saffron) on cell viability and cytokine release of stimulated peripheral blood mononuclear cells by phytohemagglutinin (PHA) and nonstimulated cells were examined. The effects of three concentrations of macerated extract, dexamethasone, and saline on cell viability and production of cytokines, including IL-4, IL-10, and IFNγ were evaluated. In cells stimulated with PHA, different concentrations of the extract significantly inhibited cell viability of lymphocytes ($P < 0.001$ for all concentrations). High concentrations of the extract (500 μg/ml) also inhibited secretion of IFNγ in stimulated cells and IL-10 secretion in both stimulated and nonstimulated cells ($P < 0.05$ for all cases). The effects of high and low concentrations of the extract (500 and 50 μg/ml, respectively) on IL-4 secretion were lower than that of dexamethasone ($P < 0.05$ to $P < 0.001$). The extract showed a stimulatory effect on IFNγ and IL-4 secretion in nonstimulated cells. The ratios of IFNγ to IL-4 in the presence of all concentrations of saffron on stimulated cells were significantly higher than for the control group ($P < 0.05$ to $P < 0.01$). These results indicated that the extract of saffron leads to increased ratio of IFNγ to IL-4 [121]. In conclusion, the results of this study showed that SE increased the ratio of IFNγ to IL-4, which indicates an increased Th1/Th2 balance. Therefore the results suggest that saffron may have therapeutic effects on inflammatory disorders that are associated with increasing Th2 cytokine production, such as asthma [121].

The mechanisms of antitumor effect were explained through the enhancing of antitumor immunity by both carotenoids and retinoids [11].

10. SAFETY

SEs as well as saffron compounds are remarkable for their lack of toxic effects. No significant hepatotoxic effect was observed in all treated groups as indicated by the constant levels of serum enzymes involved in liver function test and bilirubin [11]. The normal mouse spleen cells were insensitive to SE [13]. Hematological and biochemical studies showed complete absence of severe toxicological manifestations at the liver, kidney, or bladder [13]. The teratogenic potential of crocetin and ATRA were evaluated in frog (*Xenopus*) embryos. The data showed that crocetin is a teratogen, but far less potent than ATRA [122].

A dose-dependent co-mutagenic effect of saffron on 2-AA-induced mutagenicity was observed. The saffron component responsible for this unusual co-mutagenic effect was known as safranal. All other components were safe with no toxic, mutagenic or co-mutagenic effects [40].

In healthy volunteer: 1 month treatment 20 mg crocin tablets or placebo. General measures of health were recorded during the study such as hematological, biochemical, hormonal, and urinary parameters in pre- and post-treatment periods [123].

11. OTHER MECHANISMS

Other mechanisms of crocin action have been suggested. Crocin antagonizes the inhibitory effect of ethanol on long-term potentiation in the rat hippocampus *in vivo* and *in vitro*. Crocin (10 μM) alone did not affect synaptic potentials mediated by non-NMDA or NMDA receptors, however, it specifically antagonizes the inhibitory effect of ethanol on NMDA receptor-mediated responses in hippocampal neurons [124].

Finally, saffron-derived compounds exert a variety of effects other than anticancer effects. Obsessive-compulsive disorder (OCD) is a common psychiatric disorder defined by the presence of obsessive thoughts and repetitive compulsive actions. The nonselective serotonin (5-HT) receptor agonist mCPP is known to induce OCD-like behavior (excessive self-grooming) in rodents and exacerbate symptoms in patients with OCD [125]. Crocins (30 and 50 mg/kg, i.p.) attenuated mCPP-induced excessive self-grooming. Thus, crocins might play a role in compulsive behavior and support a functional interaction between crocins and the serotonergic system [125].

12. CONCLUSIONS

SEs as well as saffron-derived compounds, crocetin, crocin, and others exhibit remarkable antitumor effects. This is shown by various *in vitro* and *in vivo* experiments. Major effects appear to involve cell cycle arrest and apoptosis induction. In addition, antioxidative function of crocin appears to play important roles in these effects.

REFERENCES

[1] S.Z. Bathaie, S.Z. Mousavi, New applications and mechanisms of action of saffron and its important ingredients, Crit. Rev. Food Sci. 50 (2010) 761–786.

[2] S.Z. Mousavi, S.Z. Bathaie, Historical uses of saffron: Identifying potential new avenues for modern research, Avicenna J. Phytomed. 1 (2011) 57–66.

[3] E. Winterstein, J. Teleczky, On stable components of saffron I Report on picrocrocin, Helv. Chim. Acta 5 (1922) 376–381.

[4] M. Solinas, A. Cichelli, Verification of saffron quality and genuineness characteristics by HPLC analysis of color and flavor components, Ind. Aliment. (Italy) 27 (1988) 634–639.

[5] W. Rodel, M. Petrzika, Analysis of the volatile components of saffron, J. High Resolut. Chromatogr. 14 (1991) 771–774.

[6] D. Basker, M. Negbi, Uses of saffron, Econ. Bot. 37 (1983) 228–236.

[7] C.J. Gadd, The Dynasty of Agade and the Gutian Invasion, University Press, Cambridge, 1963.

[8] J.L. Gainer, D.A. Wallis, J.R. Jones, The effect of crocetin on skin papillomas and Rous sarcoma, Oncology 33 (1976) 222–224.

[9] M.M. Mathews-Roth, Effect of crocetin on experimental skin tumors in hairless mice, Oncology 39 (1982) 362–364.

[10] C.J. Wang, M.J. Lee, M.C. Chang, J.K. Lin, Inhibition of tumor promotion in benzo [a]pyrene-initiated CD-1 mouse skin by crocetin, Carcinogenesis 16 (1995) 187–191.

[11] C.J. Wang, M.Y. Chou, J.K. Lin, Inhibition of growth and development of the transplantable C-6 glioma cells inoculated in rats by retinoids and carotenoids, Cancer Lett. 48 (1989) 135–142.

[12] C.J. Wang, J.K. Lin, Inhibitory effects of carotenoids and retinoids on the in vitro growth of rat C-6 glioma cells, Proc. Natl. Sci. Counc. Repub. China B 13 (1989) 176–183.
[13] S.C. Nair, B. Pannikar, K.R. Panikkar, Antitumour activity of saffron (Crocus sativus), Cancer Lett. 57 (1991) 109–114.
[14] J. Escribano, G.L. Alonso, M. CocaPrados, J.A. Fernandez, Crocin, safranal and picrocrocin from saffron (Crocus sativus L) inhibit the growth of human cancer cells in vitro, Cancer Lett. 100 (1996) 23–30.
[15] J. Tavakkol-Afshari, A. Brook, S.H. Mousavi, Study of cytotoxic and apoptogenic properties of saffron extract in human cancer cell lines, Food Chem. Toxicol. 46 (2008) 3443–3447.
[16] S.H. Mousavi, J. Tavakkol-Afshari, A. Brook, I. Jafari-Anarkooli, Role of caspases and Bax protein in saffron-induced apoptosis in MCF-7 cells, Food Chem. Toxicol. 47 (2009) 1909–1913.
[17] S. Samarghandian, J. Tavakkol Afshari, S. Davoodi, Suppression of pulmonary tumor promotion and induction of apoptosis by Crocus sativus L. extraction, Appl. Biochem. Biotechnol. 164 (2011) 238–247.
[18] A.M. D'Alessandro, A. Mancini, A.R. Lizzi, A. De Simone, C.E. Marroccella, G.L. Gravina, et al., Crocus sativus stigma extract and its major constituent crocin possess significant antiproliferative properties against human prostate cancer, Nutr. Cancer 65 (2013) 930–942.
[19] F.I. Abdullaev, G.D. Frenkel, The effect of saffron on intracellular DNA, RNA and protein-synthesis in malignant and nonmalignant human-cells, Biofactors 4 (1992) 43–45.
[20] F.I. Abdullaev, G.D. Frenkel, Effect of saffron on cell colony formation and cellular nucleic acid and protein synthesis, Biofactors 3 (1992) 201–204.
[21] F.I. Abdullaev, Cancer chemopreventive and tumoricidal properties of saffron (Crocus sativus L.), Exp. Biol. Med. 227 (2002) 20–25.
[22] A. Bolhassani, A. Khavari, S.Z. Bathaie, Saffron and natural carotenoids: Biochemical activities and anti-tumor effects, BBA - Rev. Cancer 2014 (1845) 20–30.
[23] B. Feizzadeh, J.T. Afshari, H. Rakhshandeh, A. Rahimi, A. Brook, H. Doosti, Cytotoxic effect of saffron stigma aqueous extract on human transitional cell carcinoma and mouse fibroblast, Urol. J. 5 (2008) 161–167.
[24] S.C. Nair, C.D. Varghese, K.R. Pannikar, S.K. Kurumboor, R.K. Parathod, Effect of saffron on vitamin A levels and its antitumor activity on the growth of solid tumors in mice, Int. J. Pharmacogn. 32 (1994) 105–114.
[25] P.A. Tarantilis, H. Morjani, M. Polissiou, M. Manfait, Inhibition of growth and induction of differentiation of promyelocytic leukemia (HL-60) by carotenoids from Crocus sativus L, Anticancer Res 14 (1994) 1913–1918.
[26] S.H. Mousavi, S.A. Moallem, S. Mehri, S. Shahsavand, H. Nassirli, B. Malaekeh-Nikouei, Improvement of cytotoxic and apoptogenic properties of crocin in cancer cell lines by its nanoliposomal form, Pharm. Biol. 49 (2011) 1039–1045.
[27] S.H. Kim, J.M. Lee, S.C. Kim, C.B. Park, P.C. Lee, Proposed cytotoxic mechanisms of the saffron carotenoids crocin and crocetin on cancer cell lines, Biochem. Cell Biol. 92 (2014) 105–111.
[28] D.C. Garcia-Olmo, H.H. Riese, J. Escribano, J. Ontanon, J.A. Fernandez, M. Atienzar, et al., Effects of long-term treatment of colon adenocarcinoma with crocin, a carotenoid from saffron (Crocus sativus L.): An experimental study in the rat, Nutr. Cancer 35 (1999) 120–126.
[29] S.Y. Eid, M.Z. El-Readi, M. Wink, Carotenoids reverse multidrug resistance in cancer cells by interfering with ABC-transporters, Phytomedicine 19 (2012) 977–987.

[30] S.K. Noureini, M. Wink, Antiproliferative effects of crocin in HepG2 cells by telomerase inhibition and hTERT down-regulation, Asian Pac. J. Cancer Prev. 13 (2012) 2305–2309.
[31] S.C. Nair, S.K. Kurumboor, J.H. Hasegawa, Saffron chemoprevention in biology and medicine: A review, Cancer Biother. 10 (1995) 257–264.
[32] H. Morjani, P. Tarantilis, M. Polissiou, M. Manfait, Growth inhibition and induction of erythroid differentiation activity by crocin, dimethylcrocetine and β-carotene on K562 tumor cells, Anticancer Res 10 (1990) 1398–1406.
[33] H.H. Aung, C.Z. Wang, M. Ni, A. Fishbein, S.R. Mehendale, J.T. Xie, et al., Crocin from Crocus sativus possesses significant anti-proliferation effects on human colorectal cancer cells, Exp. Oncol. 29 (2007) 175–180.
[34] M. Rastgoo, H. Hosseinzadeh, H. Alavizadeh, A. Abbasi, Z. Ayati, M.R. Jaafari, Antitumor activity of pegylated nanoliposomes containing crocin in mice bearing C26 colon carcinoma, Planta Med. 79 (2013) 447–451.
[35] H. Bakshi, S. Sam, R. Rozati, P. Sultan, T. Islam, B. Rathore, et al., DNA fragmentation and cell cycle arrest: A hallmark of apoptosis induced by crocin from kashmiri saffron in a human pancreatic cancer cell line, Asian Pac. J. Cancer Prev. 11 (2010) 675–679.
[36] R. Hoshyar, S.Z. Bathaie, M. Sadeghizadeh, Crocin triggers the apoptosis through increasing the Bax/Bcl-2 ratio and caspase activation in human gastric adenocarcinoma, AGS, cells, DNA Cell Biol. 32 (2013) 50–57.
[37] S.Z. Bathaie, R. Hoshyar, H. Miri, M. Sadeghizadeh, Anticancer effects of crocetin in both human adenocarcinoma gastric cancer cells and rat model of gastric cancer, Biochem. Cell Biol. 91 (2013) 397–403.
[38] T. Konoshima, M. Takasaki, H. Tokuda, S. Morimoto, H. Tanaka, E. Kawata, et al., Crocin and crocetin derivatives inhibit skin tumour promotion in mice, Phytother. Res. 12 (1998) 400–404.
[39] R. Jagadeeswaran, C. Thirunavukkarasu, P. Gunasekaran, N. Ramamurty, D. Sakthisekaran, In vitro studies on the selective cytotoxic effect of crocetin and quercetin, Fitoterapia 71 (2000) 395–399.
[40] F.I. Abdullaev, L. Riveron-Negrete, H. Caballero-Ortega, J.M. Hernandez, I. Perez-Lopez, R. Pereda-Miranda, et al., Use of in vitro assays to assess the potential antigenotoxic and cytotoxic effects of saffron (Crocus sativus L.), Toxicol. In Vitro 17 (2003) 731–736.
[41] A. Dhar, S. Mehta, G. Dhar, K. Dhar, S. Banerjee, P. Van Veldhuizen, et al., Crocetin inhibits pancreatic cancer cell proliferation and tumor progression in a xenograft mouse model, Mol. Cancer Ther. 8 (2009) 315–323.
[42] S.Z. Bathaie, H. Miri, M.A. Mohagheghi, M. Mokhtari-Dizaji, A.A. Shahbazfar, H. Hasanzadeh, Saffron aqueous extract inhibits the chemically-induced gastric cancer progression in the wistar albino rat, Iranian J. Basic Med. Sci. 16 (2013) 27–38.
[43] M. Ashrafi, Mechanistic Study of the Effect of Crocin and Crocetin on NMU-Induced Breast Cancer in Rat, Tarbiat Modares University, Tehran, 2013.
[44] Y.J. Zhong, F. Shi, X.L. Zheng, Q. Wang, L. Yang, H. Sun, et al., Crocetin induces cytotoxicity and enhances vincristine-induced cancer cell death via p53-dependent and -independent mechanisms, Acta Pharmacol. Sin. 32 (2011) 1529–1536.
[45] F.I. Abdullaev, Inhibitory effect of crocetin on intracellular nucleic acid and protein synthesis in malignant cells, Toxicol. Lett. 70 (1994) 243–251.
[46] D.G. Chryssanthi, F.N. Lamari, G. Iatrou, A. Pylara, N.K. Karamanos, P. Cordopatis, Inhibition of breast cancer cell proliferation by style constituents of different Crocus species, Anticancer Res 27 (2007) 357–362.
[47] T.Z. Liu, Z.Y. Qian, Pharmacokinetics of crocetin in rats, Yao Xue Xue Bao 37 (2002) 367–369.

[48] A. Asai, T. Nakano, M. Takahashi, A. Nagao, Orally administered crocetin and crocins are absorbed into blood plasma as crocetin and its glucuronide conjugates in mice, J. Agric. Food Chem. 53 (2005) 7302–7306.
[49] L. Xi, Z. Qian, P. Du, J. Fu, Pharmacokinetic properties of crocin (crocetin digentiobiose ester) following oral administration in rats, Phytomedicine 14 (2007) 633–636.
[50] A.H. Mohammadpour, M. Ramezani, N.T. Anaraki, B. Malaekeh-Nikouei, S.A. Farzad, H. Hosseinzadeh, Development and validation of HPLC method for determination of crocetin, a constituent of saffron, in human serum samples, Iranian J. Basic Med. Sci. 16 (2013) 47–55.
[51] F. Fadai, B. Mousavi, Z. Ashtari, N. Ali Beigi, S. Farhang, S. Hashempour, et al., Saffron aqueous extract prevents metabolic syndrome in patients with schizophrenia on olanzapine treatment: A randomized triple blind placebo controlled study, Pharmacopsychiatry 47 (2014) 156–161.
[52] F.I. Abdullaev, E. Gonzalez de Mejia, Inhibition of colony formation of Hela cells by naturally occurring and synthetic agents, Biofactors 5 (1995) 133–138.
[53] S.C. Nair, M.J. Salomi, C.D. Varghese, B. Panikkar, K.R. Panikkar, Effect of saffron on thymocyte proliferation, intracellular glutathione levels and its antitumor activity, Biofactors 4 (1992) 51–54.
[54] J. Sun, X.M. Xu, C.Z. Ni, H. Zhang, X.Y. Li, C.L. Zhang, et al., Crocin inhibits proliferation and nucleic acid synthesis and induces apoptosis in the human tongue squamous cell carcinoma cell line Tca8113, Asian Pac. J. Cancer Prev. 12 (2011) 2679–2683.
[55] S.Z. Bathaie, A. Bolhasani, R. Hoshyar, B. Ranjbar, F. Sabouni, A.A. Moosavi-Movahedi, Interaction of saffron carotenoids as anticancer compounds with ctDNA, Oligo (dG.dC)15, and Oligo (dA.dT)15, DNA Cell Biol. 26 (2007) 533–540.
[56] C.D. Kanakis, P.A. Tarantilis, H.A. Tajmir-Riahi, M.G. Polissiou, Interaction of tRNA with safranal, crocetin, and dimethylcrocetin, J. Biomol. Struct. Dyn. 24 (2007) 537–546.
[57] R. Hoshyar, S.Z. Bathaie, A. Kyani, M.F. Mousavi, Is there any interaction between telomeric DNA structures, G-quadruplex and I-motif, with saffron active metabolites? Nucleosides Nucleotides Nucleic Acids 31 (2012) 801–812.
[58] W.C. Chang, Y.L. Lin, M.J. Lee, S.J. Shiow, C.J. Wang, Inhibitory effect of crocetin on benzo(a)pyrene genotoxicity and neoplastic transformation in C3H10T1/2 cells, Anticancer Res 16 (1996) 3603–3608.
[59] K. Premkumar, C. Thirunavukkarasu, S.K. Abraham, S.T. Santhiya, A. Ramesh, Protective effect of saffron (Crocus sativus L.) aqueous extract against genetic damage induced by anti-tumor agents in mice, Hum. Exp. Toxicol. 25 (2006) 79–84.
[60] H. Hosseinzadeh, A. Abootorabi, H.R. Sadeghnia, Protective effect of crocus sativus stigma extract and crocin (trans-crocin 4) on methyl methanesulfonate-induced DNA damage in mice organs, DNA Cell Biol. 27 (2008) 657–664.
[61] S.L. Willett, T.L. Miller, Binding of crocetin to plasma-albumin, Ohio J. Sci. 79 (1979) 53.
[62] T.L. Miller, S.L. Willett, M.E. Moss, J. Miller, B.A. Belinka Jr., Binding of crocetin to plasma albumin, J. Pharm. Sci. 71 (1982) 173–177.
[63] F. Zsila, Z. Bikadi, M. Simonyi, Induced chirality upon crocetin binding to human serum albumin: Origin and nature, Tetrahedron-Asymmetry 12 (2001) 3125–3137.
[64] C.D. Kanakis, P.A. Tarantilis, H.A. Tajmir-Riahi, M.G. Polissiou, Crocetin, dimethylcrocetin, and safranal bind human serum albumin: Stability and antioxidative properties, J. Agric. Food Chem. 55 (2007) 970–977.

[65] M. Ashrafi, S.Z. Bathaie, M. Taghikhani, A.A. Moosavi-Movahedi, The effect of carotenoids obtained from saffron on histone H1 structure and H1-DNA interaction, Int. J. Biol. Macromol. 36 (2005) 246–252.
[66] H. Zarei Jaliani, G.H. Riazi, S.M. Ghaffari, O. Karima, A. Rahmani, The effect of the crocus sativus l. carotenoid, crocin, on the polymerization of microtubules, in vitro, Iranian J. Basic Med. Sci. 16 (2013) 101–107.
[67] A. Ghahghaei, S.Z. Bathaie, E. Bahraminejad, Mechanisms of the effects of crocin on aggregation and deposition of a beta 1-40 fibrils in alzheimer's disease, Int. J. Pept. Res. Ther. 18 (2012) 347–351.
[68] A. Ghahghaei, S.Z. Bathaie, H. Kheirkhah, E. Bahraminejad, The protective effect of crocin on the amyloid fibril formation of a beta 42 peptide in vitro, Cell. Mol. Biol. Lett. 18 (2013) 328–339.
[69] J.H. Ahn, Y. Hu, M. Hernandez, J.R. Kim, Crocetin inhibits beta-amyloid fibrillization and stabilizes beta-amyloid oligomers, Biochem. Biophys. Res. Commun. 414 (2011) 79–83.
[70] C.F. Lv, C.L. Luo, H.Y. Ji, P. Zhao, Influence of crocin on gene expression profile of human bladder cancer cell lines T24, Zhongguo Zhong Yao Za Zhi 33 (2008) 1612–1617.
[71] P. Zhao, C.L. Luo, X.H. Wu, H.B. Hu, C.F. Lv, H.Y. Ji, Proliferation apoptotic influence of crocin on human bladder cancer T24 cell line, Zhongguo Zhong Yao Za Zhi 33 (2008) 1869–1873.
[72] X. Li, T. Huang, G. Jiang, W. Gong, H. Qian, C. Zou, Synergistic apoptotic effect of crocin and cisplatin on osteosarcoma cells via caspase induced apoptosis, Toxicol. Lett. 221 (2013) 197–204.
[73] C.J. Wang, T.C. Cheng, J.Y. Liu, F.P. Chou, M.L. Kuo, J.K. Lin, Inhibition of protein kinase C and proto-oncogene expression by crocetin in NIH/3 T3 cells, Mol. Carcinog. 17 (1996) 235–240.
[74] J.D. Hsu, F.P. Chou, M.J. Lee, H.C. Chiang, Y.L. Lin, S.J. Shiow, et al., Suppression of the TPA-induced expression of nuclear-protooncogenes in mouse epidermis by crocetin via antioxidant activity, Anticancer Res 19 (1999) 4221–4227.
[75] C.H. Zhou, M. Xiang, S.Y. He, Z.Y. Qian, Protein kinase C pathway is involved in the inhibition by crocetin of vascular smooth muscle cells proliferation, Phytother. Res. 24 (2010) 1680–1686.
[76] S. Zheng, Z. Qian, N. Wen, L. Xi, Crocetin suppresses angiotensin II-induced vascular smooth-muscle cell proliferation through inhibition of ERK1/2 activation and cell-cycle progression, J. Cardiovasc. Pharmacol. 50 (2007) 519–525.
[77] C.H. Zhou, Z.Y. Qian, M. Xiang, S.Y. He, Involvement of Ca2+ in the inhibition by crocetin of angiotensin II-induced ERK1/2 activation in vascular smooth muscle cells, Eur. J. Pharmacol. 554 (2007) 85–91.
[78] C.H. Zhou, M. Xiang, S.Y. He, Z.Y. Qian, Crocetin inhibits cell cycle G(1)/S transition through suppressing cyclin D1 and elevating p27(kip1) in vascular smooth muscle cells, Phytother. Res. 24 (2010) 975–981.
[79] C.Y. Li, W.F. Huang, Q.L. Wang, F. Wang, E. Cai, B. Hu, et al., Crocetin induces cytotoxicity in colon cancer cells via p53-independent mechanisms, Asian Pac. J. Cancer Prev. 13 (2012) 3757–3761.
[80] D.G. Chryssanthi, P. Dedes, F.N. Lamari, N. Karamanos, P. Cordopatis, Crocetin alters the gene expression pattern of metalloproteinases and their inhibitors in breast cancer cells MDA-MB-231, Planta Med. 74 (2008) 993.
[81] D.G. Chryssanthi, P.G. Dedes, N.K. Karamanos, P. Cordopatis, F.N. Lamari, Crocetin inhibits invasiveness of MDA-MB-231 breast cancer cells via downregulation of matrix metalloproteinases, Planta Med. 77 (2011) 146–151.

[82] L. Yang, Z. Qian, H. Ji, R. Yang, Y. Wang, L. Xi, et al., Inhibitory effect on protein kinase Ctheta by Crocetin attenuates palmitate-induced insulin insensitivity in 3T3-L1 adipocytes, Eur. J. Pharmacol. 642 (2010) 47–55.

[83] B.H. Wang, G.M. Polya, Selective inhibition of cyclic AMP-dependent protein kinase by amphiphilic triterpenoids and related compounds, Phytochemistry 41 (1996) 55–63.

[84] S.S. Maysam, B.S. Zahra, H. Heydarzade, Effect of crocin and crocetin on EDA activity in NMU-induced breast cancer in rat, Clin. Biochem. 44 (2011) S27.

[85] K. Bajbouj, J. Schulze-Luehrmann, S. Diermeier, A. Amin, R. Schneider-Stock, The anticancer effect of saffron in two p53 isogenic colorectal cancer cell lines, BMC Complem. Altern. Med. 12 (2012) 69.

[86] C. Kang, H. Lee, E.S. Jung, R. Seyedian, M. Jo, J. Kim, et al., Saffron (Crocus sativus L.) increases glucose uptake and insulin sensitivity in muscle cells via multipathway mechanisms, Food Chem. 135 (2012) 2350–2358.

[87] J.K. Lin, C.J. Wang, Protection of crocin dyes on the acute hepatic damage induced by aflatoxin B1 and dimethylnitrosamine in rats, Carcinogenesis 7 (1986) 595–599.

[88] C.J. Wang, J.D. Hsu, J.K. Lin, Suppression of aflatoxin B1-induced hepatotoxic lesions by crocetin (a natural carotenoid), Carcinogenesis 12 (1991) 1807–1810.

[89] C.J. Wang, H.S. Shiah, J.K. Lin, Modulatory effect of crocetin on aflatoxin B1 cytotoxicity and DNA adduct formation in C3H10T1/2 fibroblast cell, Cancer Lett. 56 (1991) 1–10.

[90] M.J. Salomi, S.C. Nair, K.R. Panikkar, Inhibitory effects of nigella-sativa and saffron (crocus-sativus) on chemical carcinogenesis in mice, Nutr. Cancer 16 (1991) 67–72.

[91] I. Das, R.N. Chakrabarty, S. Das, Saffron can prevent chemically induced skin carcinogenesis in Swiss albino mice, Asian Pac. J. Cancer Prev. 5 (2004) 70–76.

[92] I. Das, S. Das, T. Saha, Saffron suppresses oxidative stress in DMBA-induced skin carcinoma: A histopathological study, Acta Histochem. 112 (2010) 317–327.

[93] K. Premkumar, S.K. Abraham, S.T. Santhiya, A. Ramesh, Protective effects of saffron (Crocus sativus Linn.) on genotoxins-induced oxidative stress in Swiss albino mice, Phytother. Res. 17 (2003) 614–617.

[94] K. Premkumar, S.K. Abraham, S.T. Santhiya, A. Ramesh, Inhibitory effects of aqueous crude extract of Saffron (Crocus sativus L.) on chemical-induced genotoxicity in mice, Asia Pac. J. Clin. Nutr. 12 (2003) 474–476.

[95] V. Magesh, J.P.V. Singh, K. Selvendiran, G. Ekambaram, D. Sakthisekaran, Antitumour activity of crocetin in accordance to tumor incidence, antioxidant status, drug metabolizing enzymes and histopathological studies, Mol. Cell. Biochem. 287 (2006) 127–135.

[96] V. Magesh, K. DurgaBhavani, P. Senthilnathan, P. Rajendran, D. Sakthisekaran, In vivo protective effect of crocetin on benzo(a)pyrene-induced lung cancer in swiss albino mice, Phytother. Res. 23 (2009) 533–539.

[97] A. Amin, A.A. Hamza, K. Bajbouj, S.S. Ashraf, S. Daoud, Saffron: A potential candidate for a novel anticancer drug against hepatocellular carcinoma, Hepatology 54 (2011) 857–867.

[98] A. Timcheh Hariri, S.A. Moallem, M. Mahmoudi, H. Hosseinzadeh, The effect of crocin and safranal, constituents of saffron, against subacute effect of diazinon on hematological and genotoxicity indices in rats, Phytomedicine 18 (2011) 499–504.

[99] B. Ghadrdoost, A.A. Vafaei, A. Rashidy-Pour, R. Hajisoltani, A.R. Bandegi, F. Motamedi, et al., Protective effects of saffron extract and its active constituent crocin against oxidative stress and spatial learning and memory deficits induced by chronic stress in rats, Eur. J. Pharmacol. 667 (2011) 222–229.

[100] M.T. Boroushaki, Protective effects of crocin against cisplatin-induced acute renal failure and oxidative stress in rats, Nephrology 13 (2008) A30.

[101] B. Naghizadeh, S.M. Mansouri, N.V. Mashhadian, Crocin attenuates cisplatin-induced renal oxidative stress in rats, Food Chem. Toxicol. 48 (2010) 2650–2655.
[102] G. Xu, Z. Gong, W. Yu, L. Gao, S. He, Z. Qian, Increased expression ratio of Bcl-2/Bax is associated with crocin-mediated apoptosis in bovine aortic endothelial cells, Basic Clin. Pharmacol. Toxicol. 100 (2007) 31–35.
[103] S. Mehri, K. Abnous, S.H. Mousavi, V.M. Shariaty, H. Hosseinzadeh, Neuroprotective effect of crocin on acrylamide-induced cytotoxicity in PC12 cells, Cell. Mol. Neurobiol. 32 (2012) 227–235.
[104] Y.Q. Zheng, J.X. Liu, J.N. Wang, L. Xu, Effects of crocin on reperfusion-induced oxidative/nitrative injury to cerebral microvessels after global cerebral ischemia, Brain Res. 1138 (2007) 86–94.
[105] A.M. Deslauriers, A. Afkhami-Goli, A.M. Paul, R.K. Bhat, S. Acharjee, K.K. Ellestad, et al., Neuroinflammation and endoplasmic reticulum stress are coregulated by crocin to prevent demyelination and neurodegeneration, J. Immunol. 187 (2011) 4788–4799.
[106] M. Karami, S.Z. Bathaie, T. Tiraihi, M. Habibi-Rezaei, J. Arabkheradmand, S. Faghihzadeh, Crocin improved locomotor function and mechanical behavior in the rat model of contused spinal cord injury through decreasing calcitonin gene related peptide (CGRP), Phytomedicine 21 (2013) 62–67.
[107] K.N. Nam, Y.M. Park, H.J. Jung, J.Y. Lee, B.D. Min, S.U. Park, et al., Anti-inflammatory effects of crocin and crocetin in rat brain microglial cells, Eur. J. Pharmacol. 648 (2010) 110–116.
[108] R.H. Yang, L.N. Yang, X.C. Shen, W.Y. Cheng, B.H. Zhao, K.H. Ali, et al., Suppression of NF-kappa B pathway by crocetin contributes to attenuation of lipopolysaccharide-induced acute lung injury in mice, Eur. J. Pharmacol. 674 (2012) 391–396.
[109] L. Meng, L. Cui, Inhibitory effects of crocetin on high glucose-induced apoptosis in cultured human umbilical vein endothelial cells and its mechanism, Arch. Pharm. Res. 31 (2008) 357–363.
[110] S.H. Mousavi, N.Z. Tayarani, H. Parsaee, Protective effect of saffron extract and crocin on reactive oxygen species-mediated high glucose-induced toxicity in PC12 cells, Cell. Mol. Neurobiol. 30 (2010) 185–191.
[111] L. Yang, Z.Y. Qian, Y. Yang, L. Sheng, H. Ji, C.H. Zhou, et al., Involvement of Ca2+ in the inhibition by crocetin of platelet activity and thrombosis formation, J. Agric. Food Chem. 56 (2008) 9429–9433.
[112] R.M. Thushara, M. Hemshekhar, M.S. Santhosh, S. Jnaneshwari, S.C. Nayaka, S. Naveen, et al., Crocin, a dietary additive protects platelets from oxidative stress-induced apoptosis and inhibits platelet aggregation, Mol. Cell. Biochem. 373 (2013) 73–83.
[113] J. Cai, F.F. Yi, Z.Y. Bian, D.F. Shen, L. Yang, L. Yan, et al., Crocetin protects against cardiac hypertrophy by blocking MEK-ERK1/2 signalling pathway, J. Cell. Mol. Med. 13 (2009) 909–925.
[114] X. Bie, Y. Chen, X. Zheng, H. Dai, The role of crocetin in protection following cerebral contusion and in the enhancement of angiogenesis in rats, Fitoterapia 82 (2011) 997–1002.
[115] N. Umigai, J. Tanaka, K. Tsuruma, M. Shimazawa, H. Hara, Crocetin, a carotenoid derivative, inhibits VEGF-induced angiogenesis via suppression of p38 phosphorylation, Curr. Neurovasc. Res. 9 (2012) 102–109.
[116] R. Yang, K. Vernon, A. Thomas, D. Morrison, N. Qureshi, C.W. Van Way 3rd., Crocetin reduces activation of hepatic apoptotic pathways and improves survival in experimental hemorrhagic shock, JPEN J. Parenter. Enteral Nutr. 35 (2011) 107–113.

[117] M. Yamauchi, K. Tsuruma, S. Imai, T. Nakanishi, N. Umigai, M. Shimazawa, et al., Crocetin prevents retinal degeneration induced by oxidative and endoplasmic reticulum stresses via inhibition of caspase activity, Eur. J. Pharmacol. 650 (2011) 110–119.
[118] F. Ishizuka, M. Shimazawa, N. Umigai, H. Ogishima, S. Nakamura, K. Tsuruma, et al., Crocetin, a carotenoid derivative, inhibits retinal ischemic damage in mice, Eur. J. Pharmacol. 703 (2013) 1–10.
[119] A.N. Assimopoulou, Z. Sinakos, V.P. Papageorgiou, Radical scavenging activity of Crocus sativus L. extract and its bioactive constituents, Phytother. Res. 19 (2005) 997–1000.
[120] T.H. Tseng, C.Y. Chu, J.M. Huang, S.J. Shiow, C.J. Wang, Crocetin protects against oxidative damage in rat primary hepatocytes, Cancer Lett. 97 (1995) 61–67.
[121] M.H. Boskabady, S.M. Seyedhosseini Tamijani, H. Rafatpanah, A. Rezaei, A. Alavinejad, The effect of Crocus sativus extract on human lymphocytes' cytokines and T helper 2/T helper 1 balance, J. Med. Food 14 (2011) 1538–1545.
[122] G. Martin, E. Goh, A.W. Neff, Evaluation of the developmental toxicity of crocetin on Xenopus, Food Chem. Toxicol. 40 (2002) 959–964.
[123] A.H. Mohamadpour, Z. Ayati, M.R. Parizadeh, O. Rajbai, H. Hosseinzadeh, Safety evaluation of crocin (a constituent of saffron) tablets in healthy volunteers, Iranian J. Basic Med. Sci. 16 (2013) 39–46.
[124] K. Abe, M. Sugiura, Y. Shoyama, H. Saito, Crocin antagonizes ethanol inhibition of NMDA receptor-mediated responses in rat hippocampal neurons, Brain Res. 787 (1998) 132–138.
[125] G. Georgiadou, P.A. Tarantilis, N. Pitsikas, Effects of the active constituents of Crocus sativus L., crocins, in an animal model of obsessive-compulsive disorder, Neurosci. Lett. 528 (2012) 27–30.

CHAPTER FIVE

Zerumbone from Ginger (Monoterpenoid)

Yoichi Matsuo*^,1^, Hiromitsu Takeyama*

*Department of Gastroenterological Surgery, Nagoya City University Graduate School of Medical Sciences, Nagoya, Japan

[1]Corresponding author: e-mail address: matsuo@med.nagoya-cu.ac.jp

Contents

Abstract

Natural products have been regarded as safe and have been shown to have anticancer activities against a variety of cancer cell types. Zerumbone was initially isolated from the rhizome of a subtropical wild shampoo ginger (*Zingiber zerumbet* Smith) and was determined to be an anti-inflammatory agent. Carcinogenesis is a multistep process involving the transformation, survival, proliferation, invasion, and angiogenesis. The pathways associated with this process have been linked to inflammation. Zerumbone inhibits carcinogenesis via reduction of inflammatory signaling, such as nuclear factor-κB (NF-κB). In various cancers, zerumbone inhibits tumor proliferation, invasion, and angiogenesis by reducing NF-κB activity. In the near future, we should focus on both careful and accurate characterization of zerumbone and better elucidation of the molecular mechanisms involved in their actions. Furthermore, we should determine the efficacy of zerumbone by *in vivo* studies using proper animal models of cancer and demonstrate their safety and effectiveness in clinical trials.

The Enzymes, Volume 36
ISSN 1874-6047
http://dx.doi.org/10.1016/B978-0-12-802215-3.00005-7

1. INTRODUCTION

Natural products have been regarded as safe and have been shown to have anticancer activities against a variety of cancer cell types [1]. Multiple epidemiological and animal studies have shown that consumption of foods rich in fruits and vegetables decreased the occurrence of cancers [2–6,1]. Previously, many reports demonstrated that there are many natural products that have anticancer activities. For example, allicin, apigenin, berberine, butein, caffeic acid, capsaicin, catechin gallate, celastrol, curcumin, epigallocatechin gallate, fisetin, flavopiridol, gambogic acid, genistein, plumbagin, quercetin, resveratrol, sanguinarine, silibinin, sulforaphane, taxol, γ-tocotrienol, and zerumbone [1]. In this chapter, we focus on zerumbone.

Zerumbone was initially isolated from the rhizome of a subtropical wild shampoo ginger (*Zingiber zerumbet* Smith) and was determined to be an anti-inflammatory agent [7]. Previously, Yob *et al.* summarized about *Z. zerumbet* Smith [8]. They described that genus Zingiber is represented by more than 100 species that are distributed mainly in Asia. This genus of plant is confined to the tropics of Asia, Malaysia, and the Pacific Islands [9,10]. Among this genus, many scientists were interested in *Z. zerumbet* Smith because it has high medical values. *Z. zerumbet*, commonly known as the pinecone or shampoo ginger, is a perennial, tuberous root herb plant. *Z. zerumbet* can be found growing naturally in damp, shaded parts of the lowland or hill slopes. It is known by various names, for example, "Hanasyoga" (Japan), "Ghatian," and "Yaiimu" (India), "Jangli adha" (Bangladesh), "Awapuhi" (Hawaii), "Hong qiu jiang" (China), "Lempoyang" (Malaysia and Indonesia), and "Haeo dam" or "Hiao dam" (Northern Thailand) [11–14].

Zerumbone was first isolated from the rhizome of a subtropical wild shampoo ginger (*Z. zerumbet* Smith), and its molecular weight was 218.3. Presently, zerumbone has been extensively studied for its effectiveness in a broad range of biological activities including antimicrobial [15], antipyretic [16], antispasmodic, antiulcer [17], antioxidant [18], antidiabetic [19], antitumor [20], anticancer [21,22], anti-inflammatory [23,13], antinociceptive and analgesic [24,25], antiallergenic [26], antiangiogenic [27], antidipogenetic [28], antiplatelet aggregation and anticoagulant [29], and hepatoprotective effects [30]. Other studies have shown that consuming the rhizome also exhibits hypolipidemic effect by reducing intestinal cholesterol absorption, which makes it useful for treating heart diseases [31,32]. These effects of zerumbone were summarized by Rahman *et al.* [33].

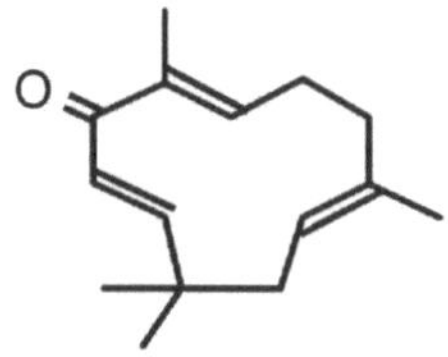

Figure 5.1 Structures of zerumbone.

2. CHARACTERISTIC FEATURE

Zerumbone is made from Zinger species. The chemical class is Sesquiterpene. The molecular formula is $C_{15}H_{22}O$, and its molecular weight is 218.3 (Fig. 5.1).

3. TARGET PATHWAYS BY ZERUMBONE

Some natural products can target survival, proliferation, invasion, angiogenesis, and metastasis steps and can influence various steps of tumor cell development by targeting one or more molecules of inflammation [1]. Zerumbone was also described to regulate these pathways.

3.1. Survival

Previous reports showed zerumbone-inhibited cell survival in various cancers.

3.1.1 Caspase Family

Abdelwahab *et al.* demonstrated that ZER has increased the number of TUNEL-positive stain and the cellular level of caspase-3, the hallmarks of apoptosis, on treated T-acute lymphoblastic leukemia, CEM-ss [34].

3.1.2 Bcl Family

Zerumbone was reported to induce the apoptotic process in liver cancer HepG2 cells through the up- and downregulation of Bax/Bcl-2 protein independent of functional p53 activity [35]. Also, Sun *et al.* described that the treatment of renal cell carcinoma (RCC) cells with zerumbone-activated caspase-3 and caspase-9, and finally led to cleavage of PARR. In addition, downregulation of Gli-1 and Bcl-2, which were closely related to the chemoresistance of RCC, was observed in zerumbone-treated RCC cells [36].

3.1.3 c-FLIP

Yodkeeree *et al.* reported that the downregulation of cFLIP by zerumbone may also lead to the enhancement of TRAIL-induced apoptosis. Zerumbone significantly downregulated the expression of cFLIP but not that of other antiapoptotic proteins [37].

3.1.4 G2/M Cell Cycle

Previous reports demonstrated that zerumone induces G2/M cell cycle arrest and apoptosis in leukemia cells and ovarian and cervical cancer cell lines [38,39].

3.2. Proliferation

3.2.1 Cyclin B1/CDK1

Cyclin B1/CDK1 correlated with cell proliferation and survival in cancer. Xian *et al.* reported that zerumbone decreased cyclin B1/CDK1 protein levels in leukemia cells [40].

3.2.2 Tumor Necrosis Factor

Zerumbone has reported to inhibit tumor necrosis factor (TNF)-α release in macrophage [41].

3.3. Invasion

The invasion is one of the important steps in cancer metastasis. Some natural products were reported to regulate the cancer invasiveness. Han previously reported that zerumbone suppressed cell migration and invasion by inhibiting MMP-3 expression in human breast cancer cells [42]. Also, ICAM-1, one of the key regulators in cancer invasiveness, was inhibitory regulated by zerumbone [43].

3.4. Angiogenesis

Angiogenesis is the formation of new capillaries from the existing vascular network and essential for tumor growth and metastasis. The progressive growth of malignant solid tumors is dependent on the development of new blood vessels that provide oxygen and nutrients to the tumor cells [44]. Angiogenesis is a complex multistep process involving extracellular matrix remodeling, endothelial cell migration and proliferation, and capillary tube formation [45]. In tumors, these steps depend on the production of angiogenic factors by both tumor and stromal cells [46]. The new blood vessels embedded in the tumor provide a gateway for the tumor cells to enter the circulation and to metastasize to distant organs, such as the liver or lung.

Thus, it is reasonable to suppose that angiogenesis is one of the first steps for tumor metastasis, such that tumor–stromal interactions with high angiogenic ability would lead to a high metastatic potential. It was previously reported that zerumbone reduced angiogenic factor from various cancers. Tsuboi *et al.* reported that zerumbone inhibited vascular endothelial growth factor (VEGF) from gastric cancer cells [47]. Also, Shamoto *et al.* demonstrated that zerumbone inhibited angiogenic factor, VEGF and interleukin-8 (IL-8) [48]. Zerumbone inhibited angiogenesis via regulation of these cytokines.

4. NUCLEAR FACTOR-KAPPA B

The inflammation is one of the major mediators of tumor progression. Because nuclear factor-kappa B (NF-κB) is one of the major inflammatory transcription factors, NF-κB is involved in the regulation of tumor cell development in various steps [1]. NF-κB can affect survival, proliferation, invasion, angiogenesis, and metastasis of the tumor. Some natural products, such as curcumin, have reported as an inhibitor of NF-κB [49]. Zerumbone also reported to inhibit inflammatory signaling, NF-κB. Especially, angiogenesis is inhibited by zerumbone by reduced NF-κB activity [47,48]. Furthermore, Sung reported that zerumbone downregulated chemokine receptor CXCR4 expression leading to inhibition of CXCL12-induced tumor cell invasion [50]. Since NF-κB-regulated cell survival, proliferation, invasion, and angiogenesis, zerumbone may have an important place in the cancer treatment (Fig. 5.2).

5. FUTURE PERSPECTIVES

Tumorigenesis is a multistep process that is regulated by various signaling pathways. Many cellular pathways overlap, which indicates a high degree of redundancy within the system. So, it is reasonable to say that to target a single molecule might ultimately have little or no effect, resulting in the need for either combination therapy or multitargeted therapy [1].

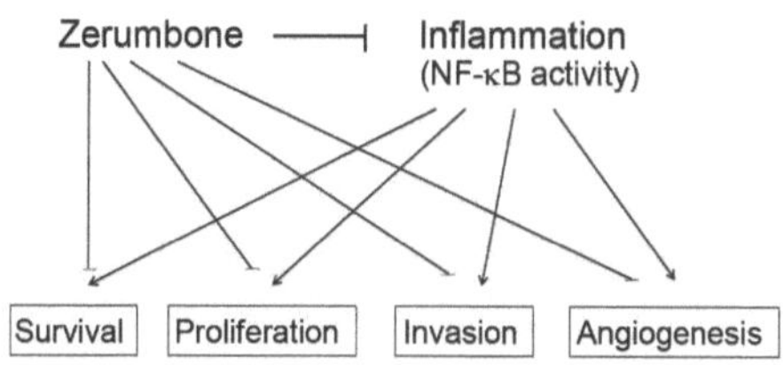

Figure 5.2 Targets of zerumbone during tumor progression.

Natural products, such as zerumbone, are inexpensive, safe, and readily available and have multitargeted potential. Some natural products have already progressed from the bench to the bedside either alone or in combination with existing therapy [1]. Moreover, we have to understand the efficacy of zerumbone in cancer prevention. In the near future, we should focus on both careful and accurate characterization of zerumbone and better elucidation of the molecular mechanisms involved in their actions. Furthermore, we should determine the efficacy of zerumbone by *in vivo* studies using proper animal models of cancer and demonstrate their safety and effectiveness in clinical trials.

REFERENCES

[1] S.C. Gupta, J.H. Kim, S. Prasad, B.B. Aggarwal, Regulation of survival, proliferation, invasion, angiogenesis, and metastasis of tumor cells through modulation of inflammatory pathways by nutraceuticals, Cancer Metastasis Rev. 29 (3) (2010) 405–434.
[2] L. Reddy, B. Odhav, K.D. Bhoola, Natural products for cancer prevention: A global perspective, Pharmacol. Ther. 99 (2003) 1–13.
[3] G. Block, B. Patterson, A. Subar, Fruit, vegetables, and cancer prevention: A review of the epidemiological evidence, Nutr. Cancer 18 (1992) 1–29.
[4] V. Benetou, P. Orfanos, P. Lagiou, D. Trichopoulos, P. Boffetta, A. Trichopoulou, Vegetables and fruits in relation to cancer risk: Evidence from the Greek EPIC cohort study, Cancer Epidemiol. Biomark. Prev. 17 (2008) 387–392.
[5] N.D. Freedman, Y. Park, A.F. Subar, A.R. Hollenbeck, M.F. Leitzmann, A. Schatzkin, et al., Fruit and vegetable intake and head and neck cancer risk in a large United States prospective cohort study, Int. J. Cancer 122 (2008) 2330–2336.
[6] K.A. Steinmetz, J.D. Potter, Vegetables, fruit, and cancer prevention: A review, J. Am. Diet. Assoc. 96 (1996) 1027–1039.
[7] A. Murakami, D. Takahashi, T. Kinoshita, et al., Zerumbone, a Southeast Asian ginger sesquiterpene, markedly suppress free radical generation, proinflammatory protein production, and cancer cell proliferation accompanied by apoptosis: The α, β-unsaturated carbonyl group is a prerequisite, Carcinogenesis 23 (2002) 795–802.
[8] N.J. Yob, S.M. Jofrry, M.M. Affandi, L.K. Teh, M.Z. Salleh, Z.A. Zakaria, *Zingiber zerumbet* (L.) Smith: A review of its ethnomedicinal, chemical, and pharmacological uses, Evid. Based Complement. Alternat. Med. 2011 (2011) 543216.
[9] I. Burkill, A Dictionary of the Economic Products of the Malay Peninsula, Ministry of Agriculture and Cooperative, Kuala Lumpur, Malaysia, 1966.
[10] B. Jantan, M.S. Mohd Yassin, C.B. Chin, L.L. Chen, N.L. Sim, Antifungal activity of the essential oils of nine Zingiberaceae species, Pharm. Biol. 41 (5) (2003) 392–397.
[11] G.C. Huang, T.Y. Chien, L.G. Chen, C.C. Wang, Antitumor effects of zerumbone from *Zingiber zerumbet* in P-388D cells in vitro and in vivo, Planta Med. 71 (3) (2005) 219–224.
[12] N.I. Bhuiyan, J.U. Chowdhury, J. Begum, Chemical investigation of the leaf and rhizome essential oils of *Zingiber zerumbet* (L.) Smith from Bangladesh, Bangladesh J. Pharmacol. 4 (1) (2009) 9–12.
[13] Z.A. Zakaria, A.S. Mohamad, C.T. Chear, Y.Y. Wong, D.A. Israf, M.R. Sulaiman, Antiinflammatory and antinociceptive activities of *Zingiber zerumbet* methanol extract in experimental model systems, Med. Princ. Pract. 19 (4) (2010) 287–294.

[14] Tushar, S. Basak, G.C. Sarma, L. Rangan, Ethnomedical uses of Zingiberaceous plants of Northeast India, J. Ethnopharmacol. 132 (1) (2010) 286–296.
[15] G. Kader, F. Nikkon, M.A. Rashid, T. Yeasmin, Antimicrobial activities of the rhizome extract of *Zingiber zerumbet* Linn, Asian Pac. J. Trop. Biomed. 1 (5) (2011) 409–412.
[16] M.N. Somchit, M.H. Shukriyah, A.A. Bustamam, A. Zuraini, Anti-pyretic and analgesic activity of *Zingiber zerumbet*, Int. J. Pharmacol. 1 (3) (2005) 277–280.
[17] M. Al-Amin, G.N.N. Sultana, C.F. Hossain, Antiulcer principle from Zingiber montanum, J. Ethnopharmacol. 141 (1) (2012) 57–60.
[18] M. Habsah, M. Amran, M.M. Mackeen, et al., Screening of Zingiberaceae extracts for antimicrobial and antioxidant activities, J. Ethnopharmacol. 72 (3) (2000) 403–410.
[19] T.-F. Tzeng, S.-S. Liou, C.J. Chang, I.-M. Liu, The ethanol extract of *Zingiber zerumbet* attenuates streptozotocin-induced diabetic nephropathy in rats, Evid. Based Complement. Alternat. Med. 2013 (2013), Article ID 340645, 8 pages.
[20] S.I.A. Wahab, A.B. Abdul, H.C. Yeel, A.S. Alzubain, M.M. Elhassan, M.M. Syam, Anti-tumor activities of analogues derived from the bioactive compound of *Zingiber zerumbet*, Int. J. Cancer Res. 4 (4) (2008) 154–159.
[21] R.A. Rashid, A.H.L. Pihie, The antiproliferative effects of *Zingiber zerumbet* extracts and fractions on the growth of human breast carcinoma cell lines, Malaysian J. Pharm. Sci. 3 (1) (2005) 45–52.
[22] A. Rasedee, S.R. Heshu, A. Ahmad Bustamam, C.W. How, K.Y. Swee, A composition for treating leukaemia, Malaysian Patent Application, p. PI2013700213 (2013).
[23] M.R. Sulaiman, E.K. Perimal, M.N. Akhtar, et al., Anti-inflammatory effect of zerumbone on acute and chronic inflammation models in mice, Fitoterapia 81 (7) (2010) 855–858.
[24] M.N. Somchit, J.H. Mak, A. Ahmad Bustamam, et al., Zerumbone isolated from *Zingiber zerumbet* inhibits inflammation and pain in rats, J. Med. Plants Res. 6 (2012) 177–180.
[25] M.R. Sulaiman, T.A.S. Tengku Mohamad, W.M. Shaik Mossadeq, et al., Antinociceptive activity of the essential oil of *Zingiber zerumbet*, Planta Med. 76 (2) (2010) 107–112.
[26] S. Tewtrakul, S. Subhadhirasakul, Anti-allergic activity of some selected plants in the Zingiberaceae family, J. Ethnopharmacol. 109 (3) (2007) 535–538.
[27] J. Rhode, S. Fogoros, S. Zick, et al., Ginger inhibits cell growth and modulates angiogenic factors in ovarian cancer cells, BMC Complement. Altern. Med. 7 (44) (2007).
[28] T.-F. Tzeng, I.-M. Liu, 6-Gingerol prevents adipogenesis and the accumulation of cytoplasmic lipid droplets in 3T3-L1 cells, Phytomedicine 20 (6) (2013) 481–487.
[29] I. Jantan, S.M. Raweh, H.M. Sirat, et al., Inhibitory effect of compounds from Zingiberaceae species on human platelet aggregation, Phytomedicine 15 (4) (2008) 306–309.
[30] A.S. El-Sharaky, A.A. Newairy, M.A. Kamel, S.M. Eweda, Protective effect of ginger extract against bromobenzene-induced hepatotoxicity in male rats, Food Chem. Toxicol. 47 (7) (2009) 1584–1590.
[31] I. Sharma, D. Gusain, V.P. Dixit, Hypolipidemic and anti-atherosclerotic effect of Zingiber officinale in cholesterol fed rabbits, Phytother. Res. 10 (1996) 517–518.
[32] U. Bhandari, J.N. Sharma, R. Zafar, The protective action of ethanolic ginger (Zingiber officinale) extract in cholesterol fed rabbits, J. Ethnopharmacol. 61 (2) (1998) 167–171.
[33] H.S. Rahman, A. Rasedee, S.K. Yeap, H.H. Othman, M.S. Chartrand, F. Namver, A.B. Abdul, How CW biomedical properties of a natural dietary plant metabolite, zerumbone, in cancer therapy and chemoprevention trials, Biomed. Res. Int. 2014 (2014) 920742, Epub 2014 Jun 16.
[34] S.I. Abdelwahab, A.B. Abdul, S. Mohan, M.M. Taha, S. Syam, M.Y. Ibrahim, A.A. Mariod, Zerumbone induces apoptosis in T-acute lymphoblastic leukemia cells, Leuk. Res. 35 (2) (2011) 268–271.

[35] S.A. Sakinah, S.T. Handayani, L.P. Hawariah, Zerumbone induced apoptosis in liver cancer cells via modulation of Bax/Bcl-2 ratio, Cancer Cell Int. 3; 7 (2007) 4.
[36] Y. Sun, Q. Sheng, Y. Cheng, Y. Xu, Y. Han, J. Wang, L. Shi, H. Zhao, C. Du, Zerumbone induces apoptosis in human renal cell carcinoma via Gli-1/Bcl-2 pathway, Pharmazie 68 (2013) 141–145.
[37] S. Yodkeeree, B. Sung, P. Limtrakul, B.B. Aggarwal, Zerumbone enhances TRAIL-induced apoptosis through the induction of death receptors in human colon cancer cells: Evidence for an essential role of reactive oxygen species, Cancer Res. 69 (16) (2009) 6581–6589.
[38] S.I. Abdelwahab, A.B. Abdul, Z.N. Zain, A.H. Hadi, Zerumbone inhibits interleukin-6 and induces apoptosis and cell cycle arrest in ovarian and cervical cancer cells, Int. Immunopharmacol. 12 (4) (2012) 594–602.
[39] H.S. Rahman, A. Rasedee, A.B. Abdul, N.A. Zeenathul, H.H. Othman, S.K. Yeap, C.W. How, W.A. Hafiza, Zerumbone-loaded nanostructured lipid carrier induces G2/M cell cycle arrest and apoptosis via mitochondrial pathway in a human lymphoblastic leukemia cell line, Int. J. Nanomedicine 9 (2014) 527–538.
[40] M. Xian, K. Ito, T. Nakazato, et al., Zerumbone, a bioactive sesquiterpene, induces G2/M cell cycle arrest and apoptosis in leukemia cells via a Fas- and mitochondria-mediated pathway, Cancer Sci. 98 (1) (2007) 118–126.
[41] H.S. Rahman, A. Rasedee, C.W. How, et al., Zerumbone-loaded nanostructured lipid carriers: Preparation, characterization, and antileukemic effect, Int. J. Nanomedicine 8 (2013) 2769–2781.
[42] J. Han, S.Y. Bae, S.J. Oh, J. Lee, J.H. Lee, H.C. Lee, S.K. Lee, W.H. Kil, S.W. Kim, S.J. Nam, S. Kim, J.E. Lee, Zerumbone suppresses IL-1β-induced cell migration and invasion by inhibiting IL-8 and MMP-3 expression in human triple-negative breast cancer cells. Phytother. Res. (2014 May 30), http://dx.doi.org/10.1002/ptr.5178. [Epub ahead of print].
[43] Y. Takada, A. Murakami, B.B. Aggarwal, Zerumbone abolishes NF-kappaB and IkappaBalpha kinase activation leading to suppression of antiapoptotic and metastatic gene expression, upregulation of apoptosis, and downregulation of invasion, Oncogene 24 (46) (2005) 6957–6969.
[44] L.A. Liotta, E.C. Kohn, The microenvironment of the tumour-host interface, Nature 411 (6835) (2001) 375–379.
[45] C.H. Blood, B.R. Zetter, Tumor interactions with the vasculature: Angiogenesis and tumor metastasis, Biochim. Biophys. Acta 1032 (1) (1990) 89–118.
[46] J. Folkman, What is the evidence that tumors are angiogenesis dependent? J. Natl. Cancer Inst. 82 (1) (1990) 4–6.
[47] K. Tsuboi, Y. Matsuo, T. Shamoto, T. Shibata, S. Koide, M. Morimoto, S. Guha, B. Sung, B.B. Aggarwal, H. Takahashi, H. Takeyama, Zerumbone inhibits tumor angiogenesis via NF-κB in gastric cancer, Oncol. Rep. 31 (1) (2014) 57–64.
[48] T. Shamoto, Y. Matsuo, T. Shibata, K. Tsuboi, T. Nagasaki, H. Takahashi, H. Funahashi, Y. Okada, H. Takeyama, Zerumbone inhibits angiogenesis by blocking NF-κB activity in pancreatic cancer, Pancreas 43 (3) (2014) 396–404.
[49] S. Shishodia, H.M. Amin, R. Lai, B.B. Aggarwal, Curcumin (diferuloylmethane) inhibits constitutive NF-kappaB activation, induces G1/S arrest, suppresses proliferation, and induces apoptosis in mantle cell lymphoma, Biochem. Pharmacol. 70 (5) (2005) 700–713.
[50] B. Sung, S. Jhurani, K.S. Ahn, Y. Mastuo, T. Yi, S. Guha, M. Liu, B.B. Aggarwal, Zerumbone down-regulates chemokine receptor CXCR4 expression leading to inhibition of CXCL12-induced invasion of breast and pancreatic tumor cells, Cancer Res. 68 (21) (2008) 8938–8944.

CHAPTER SIX

Research Progress on Natural Triterpenoid Saponins in the Chemoprevention and Chemotherapy of Cancer

Jun-Rong Du*,[1], Fang-Yi Long*, Chu Chen†,[1]
*Department of Pharmacology, West China School of Pharmacy, Sichuan University, Chengdu, P.R. China
†Institute of Pharmaceutical Research, Sichuan Academy of Chinese Medicine Sciences, Chengdu, P.R. China
[1]Corresponding authors: e-mail address: dujr_1@163.com; chenchu1978@qq.com

Contents

Abstract

Triterpenoid saponins are glycosides with remarkable structural and bioactive diversity. They are becoming increasingly significant in the treatment of cancer due to their efficacy and safety. This chapter provides an update on the sources, pharmacological effects, structure–activity relationships, and clinical studies of anticancer triterpenoid saponins with a particular focus on the molecular mechanisms underlying their therapeutic properties. The correlative references and study reports described were collected through PubMed. The anticancer triterpenoid saponins enable the inhibition of cancer formation and progression by modulating multiple signaling targets related to cellular proliferation, apoptosis, autophagy, metastasis, angiogenesis, inflammation, oxidative

The Enzymes, Volume 36
ISSN 1874-6047
http://dx.doi.org/10.1016/B978-0-12-802215-3.00006-9

stress, multidrug resistance, cancer stem cells, and microRNAs. This review provides new insights into the molecular basis of triterpenoid saponins in the chemoprevention and chemotherapy of cancer.

1. INTRODUCTION

Triterpenoids are composed of 30 carbon atoms, polymerized to form six isoprene units. Triterpenoids are biosynthesized by the cyclization of its precursor squalene, a hydrocarbon composed of two molecules of farnesyl pyrophosphates joined tail to tail [1]. Based on their chemical structures, triterpenoids can be grouped into linear, monocyclic, dicyclic, etc., up through pentacyclic compounds.

Triterpenoids are widely distributed in nature; they may be found in fungi, ferns, higher plants, animals, and marine organisms. *In vivo*, triterpenoids have been found in both free states and in combination with sugars forming glycosides and esters. Specifically, triterpenoid saponins are glycosides consisting of a sugar moiety (glycone) and a triterpenoid component (aglycone). They are generally water soluble and characterized by their strong foam-forming and surfactant properties in aqueous solution [2]. The word saponin comes from the Latin *sapo*, meaning soap. Plant materials with significant amounts of saponins have traditionally been used as detergents, for example, soapbark (*Quillaja saponaria*; Rosaceae), soapwort (*Saponaria officinalis*; Caryophyllaceae), and shikakai powder (*Gleditsia sinensis*; Leguminosae). Although these materials have hemolytic properties and are thus highly toxic when injected into the blood, saponins are relatively harmless under oral administration as they are hydrolyzed within the gastrointestinal tract and have low absorption [1]. Though triterpenoid saponins are found rarely in monocotyledons, they are abundant in many dicotyledonous families such as Araliaceae, Ranunculaceae, Leguminosae, Caryophyllaceae, Cucurbitaceae, and Umbelliferae [3]. The aglycone component of saponin is termed sapogenin and it can be classified according to its skeletal organization. The most common types are tetracyclic triterpenoids (cucurbitane, dammarane, and lanostane) and pentacyclic triterpenoids (lupine and oleanane) (Fig. 6.1). The glycone component of saponin is an oligosaccharide which may be arranged in a linear, branched, or cyclical fashion, and whose frequent sugars are D-glucose, D-xylose, D-galactose, D-glucuronic acid, D-galacturonic acid, L-rhamnose, and L-arabinose. Generally, the glycone and aglycone components of triterpenoid saponins are connected by ether linkages; though ester linkages occur, they are relatively rare. In

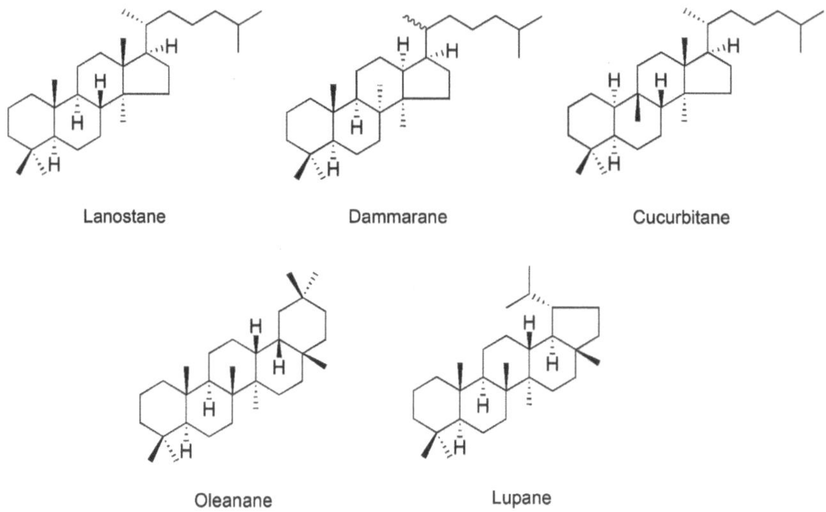

Figure 6.1 The common structural types of triterpenoids.

some cases, however, the two types of linkages have been found to coexist within a single molecule [4].

A significant amount of triterpenoid saponins are found in foods such as beans, soybeans, spinach, lentils, and oats. Moreover, they may be obtained from many valuable phytomedicines, such as Radix et Rhizoma Ginseng, Radix et rhizoma glycyrrhizae, and Radix astragli [1,4,5]. These medicines are traditionally used in the treatment of a number of diseases including cancer, diabetes, cardiac-cerebral vascular diseases, inflammation, and viral and bacterial infections.

Conventional anticancer agents are highly toxic not only to tumor cells but also to normal cells. Therefore, novel natural anticancer compounds provide an attractive alternative to synthetic compounds due to their favorable efficacy and safety profiles. The objective of this review is to provide an update on the sources, pharmacological effects, structure–activity relationship, and clinical studies of anticancer triterpenoid saponins with a particular focus on the molecular mechanisms underlying their anticancer properties.

2. TRITERPENOID SAPONINS IN THE PREVENTION AND THERAPY OF CANCERS

Natural products are an excellent source of biologically active compounds. It is well known that a large proportion of modern drugs have been developed from natural products or their synthetic analogues. Some phytochemicals in herbal medicine exert anticancer effects and are beneficial for the treatment and survivability of cancer patients. In recent years, many natural triterpenoid saponins have been found to have significant chemopreventive and chemotherapeutic effects in a wide spectrum of malignant cells *in vitro* and *in vivo*. The anticancer properties and potential mechanisms of action of natural triterpenoid saponins are summarized in Table 6.1 [6–97].

As shown in Table 6.1, the cytotoxicity of triterpenoid saponins from plants have been extensively screened in a variety of cancer cells *in vitro*, including human and/or murine cancer cell lines of bile duct, bladder, blood, brain, breast, cervix, chorion, colon, esophagus, liver, pancreas, prostate, stomach, ovary, oral epithelia, skin, and sarcoma. In particular, several triterpenoid saponins have been observed to exert higher sensitivity on cancer cells than normal cells, suggesting their potential safety as anticancer agents [6,15,28,38,44,45,53,67,74,81,86,91]. Furthermore, the anticancer activities of triterpenoid saponins have been evaluated in various animal models bearing human or animal cancers [18,21,33,34,41,42,54,56,58, 60,64–67,77,78,80–82,89–92,97]. Of note, the tested triterpenoid saponins were generally very well tolerated in animals at doses that caused tumor regression, and showed no observable toxicity, such as injury of organs or changes in body weight.

3. ANTICANCER PROPERTIES AND MOLECULAR MECHANISMS OF TRITERPENOID SAPONINS

With growing insight into the mechanisms underlying the intrinsic cellular and extrinsic microenvironmental factors contributing to cancer progression, a number of molecular targets have been identified for the development of the novel anticancer agents (Fig. 6.2). In recent years, a variety of natural triterpenoid saponins have been found to have anticancer properties by modulating the specific targets of cancer cells. Below, we elaborate on some of these anticancer properties highlighting the molecular mechanisms of triterpenoid saponins.

3.1. Inhibition of Proliferation

Deregulation of cellular proliferation is one of the most frequent alterations associated with carcinogenesis. Therefore, inhibiting cellular proliferation is a feasible approach for targeting cancer cells. In this regard, a number of studies have shown antiproliferative properties of various triterpenoid saponins such as astragaloside IV [24], β-escin (escin, aescin) [17], Echinoside A [41],

Table 6.1 Anticancer Properties and Potential Mechanisms of Triterpenoid Saponins that Have Been Reported in Either *In Vitro* or *In Vivo* Studies

Species	Saponins	Kinds of Cancer	Molecular Mechanisms	References
Acacia victoriae	Avicins D and G	Cells: Jurkat, NB4, U2OS	Activates mitochondria and Fas-caspase pathways	[6,7]
	Avicin D	Cells: A549, MDA-MB-231, PC3, SKOV3, T47D, U2OS	AMPK activation	[8]
	Avicin D	Cells: HH, Jurkat, MJ	Inhibits Hsp70 and XIAP proteins	[9]
	Avicin D	Cells: U266	Inhibits STAT3 signaling	[10]
Achyranthes fauriei	Achyranthoside H methyl ester	Cells: MCF-7, MDA-MB-453	Induces cleavage of PARP	[11]
Aesculus hippocastanum	Escin	Cells: A293, H1299, KBM-5, Jurkat, U266	Inhibits NF-κB signaling	[12]
	β-Escin	Cells: MZ-ChA-1, QBC939, SK-ChA-1, SMMC-7721	Activates intrinsic and extrinsic caspase pathways	[13,14]
	β-Escin	Cells: A549, HepG2	Inhibits JAK/STAT signaling and p38	[15,16]
	β-Escin	Cells: HT-29	Induces $p21^{waf1/cip1}$ expression	[17]
	β-Escin	Cells: H460; Mice: tobacco carcinogen induced lung cancer	Inhibits expression of ALDH1A1 and RhoA/Rock	[18]

Continued

Table 6.1 Anticancer Properties and Potential Mechanisms of Triterpenoid Saponins that Have Been Reported in Either *In Vitro* or *In Vivo* Studies—cont'd

Species	Saponins	Kinds of Cancer	Molecular Mechanisms	References
Androsace umbellata	Saxifragifolin D	Cells: MCF-7, MDA-MB-231	Induces ROS-mediated endoplasmic reticulum stress	[19]
Anemone raddeana	Raddeanin A	Cells: BGC-823, MKN-28, SGC-7901	Induces invasion-related protein expression and caspase activation	[20]
Ardisia pusilla	Ardipusilloside I	Cells and mice: HepG2, SMMC-7721	Increases E-cadherin via Rac1 activation and inhibits MMPs and Mek/Erk and Akt signaling	[21]
Astragalus membranaceus var. *mongolicus*	Astragalus saponins	Cells: AGS	Activates caspase-3 and reduces cyclin B1, MMPs and VEGF	[22]
Astragalus membranaceus	Astragaloside II	Cells: Bel-7402, Bel-7402/FU	Inhibits MAPKs and reduces P-gp and *MDR1* expression	[23]
Astragalus sp.	Astragaloside IV	Cells: HepG2	Inhibits Vav3.1 expression	[24]
Bolbostemma paniculatum	Tubeimoside-1	Cells: A549, Bel-7402, CNE-2Z, Eca-109, Hela, HepG2, JEG-3, PC-3, SGC-7901	Induces intrinsic and extrinsic caspase pathways and NF-κB signaling	[25–29]
	Tubeimoside-1	Cells: SKOV-3	Induces GRP78/Bip and CHOP expression and intrinsic caspase pathway	[30]

Bupleurum sp.	Saikosaponin a	Cells: HCT116, LoVo, SW48, SW480	Activates caspase-2 and -8	[31]
Bupleurum falcatum	Saikosaponin-D	Cells: HeLa, MCF-7	Inhibits SERCA	[32]
	Saikosaponin-D	Diethylnitrosamine induced rat hepatoma	Inhibits C/EBPβ and COX-2	[33]
	Saikosaponin-D	N-diethylnitrosamine induced rat hepatoma	Reduces syndecan-2, MMP-2 and -13	[34]
Cimicifuga heracleifolia	Cimiside E	Cells: AGS	Activates intrinsic and extrinsic caspase pathways	[35]
Corchorus acutangulus	Corchorusin-D	Cells: K562	Activates intrinsic and extrinsic caspases and Akt/PKB pathways	[36]
Dipsacus asperoides	Akebia saponin PA	Cells: AGS, MKN-45, SNU-638, KATO III	Modulates Akt/mTOR and AMPK/mTOR	[37]
Entada phaseoloides	Phaseoloideside E	Cells: Ec-109	Activates mitochondrial caspase pathway	[38]
Glochidion eriocarpum	Glochieriosides A-D	Cells: HL-60, HCT-116	Activates mitochondrial caspase pathway and ERK and p38	[39]
Glycine sp.	Soyasaponin Bb	Cells: Eca-9706	Inhibits NF-κB/HDAC1 and activates PETEN and caspase-3	[40]
Holothuria nobilis	Echinoside A	Cells: various cell lines; Mice: H22, PC-3, S180	Inhibits Top2α binding to DNA	[41]

Continued

Table 6.1 Anticancer Properties and Potential Mechanisms of Triterpenoid Saponins that Have Been Reported in Either *In Vitro* or *In Vivo* Studies—cont'd

Species	Saponins	Kinds of Cancer	Molecular Mechanisms	References
Kalopanax pictus	Kalopanaxsaponin-A	Cells and mice: YD-10B	Reduces MMP-9 via HuR and Rab1A	[42]
Lonicera macranthoides	Macranthoside B	Cells: HL-60, HepG2, Jurkat, K-562, LoVo, U-937	Caspase-3 activation	[43]
Lysimachia ciliata	Desglucoanagalloside B	Cells: DU-145, PC-3	Activates mitochondrial caspase pathway	[44]
Panax ginseng	Ginsenoside F2	Cells: breast cancer stem	Activates p53 and mitochondrial caspase pathway	[45]
	Protopanaxatriol ginsenosides	Cells: AML-2/D100, AML-2/DX100	Inhibits P-gp function	[46]
	20(*S*)-Protopanaxatriol, Ginsenosides Rb1and Rg1	Cells: HepG2-C8, TRAMP C1	Activates Nrf2-ARE pathway	[47]
	Ginsenoside Rd	Cells: MCF-7, MCF-7/ADR	Induces P-gp degradation	[48]
	Ginsenoside Rd	Cells: HepG2	Reduces MMPs via MAPKs inactivation and EMT	[49]
	Ginsenoside Rg1	Cells: K562	Enhances p16, p53, p21, and Rb	[50]
	Ginsenoside Rg1	Cells: TF-1/Epo	Inhibits EpoR-mediated apoptosis pathway	[51]

	Ginsenoside Rg3	Cells: DU145, HCT116, LNCaP, PC-3, SW620	Inhibits NF-κB signaling	[52,53]
	Gensenoside Rg3	Cells: Eca-109, 786-0; Mice: Lewis lung carcinoma	Inhibits VEGF signaling	[54-56]
	Ginsenoside Rg3	Cells: Hep1-6, HepG2, SMMC-7721 Mice: Hep1-6	Activates mitochondrial caspase cascade	[57,58]
	Ginsenoside Rg3	Cells: AGS	Blocks TRPM7	[59]
	Ginsenoside Rg3	Cells: CT-26	Scavenges ROS	[60]
	Ginsenoside Rg3	Cells: PANC-1	Decreases Pim-3 and p-Bad	[61]
	Ginsenoside Rg3	Cells: PC-3 M	Reduces aquaporin 1 expression	[62]
	Ginsenoside Rg3	Cells: SK-MES-1	Inhibits survivin expression	[63]
	Ginsenoside Rg3	Cells and mice: HCT116	Inhibits Wnt/β-catenin signaling	[64]
	Ginsenoside Rg3	Cells: KB, KBV20C; Mice: P388/DOX	Inhibits P-gp function	[65]
	20(*S*)-Ginsenoside Rg3	Cells: Caco-2; Mice: MCF-7	Decreases P-gp expression	[66]
	20(*S*)-Ginsenoside Rg3	Cells: HepG2, Hep3B, SK-Hep1, Huh-7; Mice: Huh-7	Induces DR5 upregulation via CHOP	[67]

Continued

Table 6.1 Anticancer Properties and Potential Mechanisms of Triterpenoid Saponins that Have Been Reported in Either *In Vitro* or *In Vivo* Studies—cont'd

Species	Saponins	Kinds of Cancer	Molecular Mechanisms	References
	Ginsenoside Rh1	Cells: HepG2	Reduce MMP-1 expression via inhibiting MAPKs	[68]
	Ginsenoside Rh2	Cells: A172, A549, T98MG, U251, U281	Decreases miR-424 and increases miR-148a, miR-128	[69,70]
	Ginsenoside Rh2	Cells: DU-145, HCT116, HeLa, HepG2	None particular mechanism reported	[71]
	Ginsenoside Rh2	Cells: HL-60, U937	Increases TGF-β	[72]
	Ginsenoside Rh2	Cells: MCF-7, MDA-MB-231	Induces binding of $p15^{Ink4B}$ and $p27^{Kip1}$	[73]
	Ginsenoside Rh2	Cells: A431	Lipid rafts internalization and Akt inactivation	[74]
	Ginsenoside Rh2	Cells: A549	Alters cell cycle regulatory proteins and increases DR4-mediated caspase pathway	[75]
	Ginsenoside Rh2	Cells: Hela	Induces Fas aggregation in lipid rafts	[76]
	Ginsenoside Rh2	Cells: 9 L, B16, LNCaP, MCF-7, MDA435/LCC6, MMRU, SF120, Skmel-110; Mice: B16, H22, S180	Caspase activation	[77]

	Ginsenoside Rh2	Cells: MCF-7, MDA-MB-231; Mice: MDA-MB-231	Activates mitochondria-mediated apoptosis	[78]
	20(*S*)-Ginsenoside Rh2	Cells: MCF-7/ADR	Inhibits P-gp function	[79]
	Protopanaxadiol	Cells: NCI-H23, PC-3, ACHN, Caki-1, K562; Mice: K562	Activates neutral sphingomyelinase 2	[80]
	20(*R*)-25-methoxyl-dammarane-3β,12β,20-triol	Cells and mice: A549, H292	p38 activation	[81]
Panax notoginseng	20(*S*)-25-methoxyl-dammarane-3β,12β,20-triol	Cells: MCF-7, MDA-MB-231 Mice: MDA-MB-231	Decreases MDM2 expression	[82]
Pearsonothuria graeffei	Ds-echinoside A	Cells: HepG2	Inhibits NF-κB-dependent MMP-9 and VEGF expressions/inhibits NF-κB signaling	[83]
Platycodon grandiflorum	Platycodin D, 2″-*O*-acetylplatycodin D, 3″-*O*-acetylplatycodin D, Polygalacin D, 2″-*O*-acetylpolygalacin D, 3″-*O*-acetylpolygalacin D	Cells: AGS, A549, Caco-2, Hela, HepG2, Jurkat, MCF-7	Activates intrinsic and extrinsic caspase pathways and inhibits Akt/mTOM and NF-κB pathways	[84]

Continued

Table 6.1 Anticancer Properties and Potential Mechanisms of Triterpenoid Saponins that Have Been Reported in Either *In Vitro* or *In Vivo* Studies—cont'd

Species	Saponins	Kinds of Cancer	Molecular Mechanisms	References
	Platycodin D	Cells: K562, THP-1, U937	Disrupts microtubule and activates caspase-3	[85]
	Platycodin D	Cells: 4 T1,MDA-MB-231	Inhibits TGF-β/Smads signaling	[86]
	Platycodin D	Cells: HepG2	Inhibits PI3K/Akt pathway	[87]
	Platycodin D	Cells: MCF-7	Activates MAPKs and intrinsic and extrinsic caspase pathways	[88]
	Platycodin D	Cells and mice: MDA-MB-231	Inhibits EGFR-mediated Akt and MAPK pathways	[89]
Polygala senega	Senegin III	Mice: S180	Increases PEDF expression	[90]
Pulsatilla koreana	Pulsatilla saponin D	Cells: AGS, MKN-28, MKN-45; Mice: MKN-45	Inhibits c-Met signaling	[91]
	Pulsatilla saponin D	Cells: HT-29, LoVo; Mice: HT-29 cells	Inhibits Akt/mTOR pathway	[92]
Salicornia bigelovii	Bigelovii A	HL-60 cells	Activates mitochondrial caspase pathway	[93]
Sanguisorba officinalis	Ziyuglycoside II	Cells: MDA-MB-435	Induces p53/p21 expression and mitochondrial caspase pathway	[94]

Trichosanthes kirilowii	Cucurbitacins-B and -I	Cells: ARO, HeLa, MCF-7, HCT-116, TPC-1 MEFs, U87	Induces ROS-mediated ERK and JNK activation	[95]
	Cucurbitacin-I	Cells: BJAB, I-83 NALM-6, primary B-CLL	Inhibits STAT3 signaling pathway	[96]
Zizyphus jujube var. *spinosa*	Jujuboside B	Cells: AGS, HCT 116; Mice: HCT 116	Induces p38/JNK activation and FasL expression	[97]

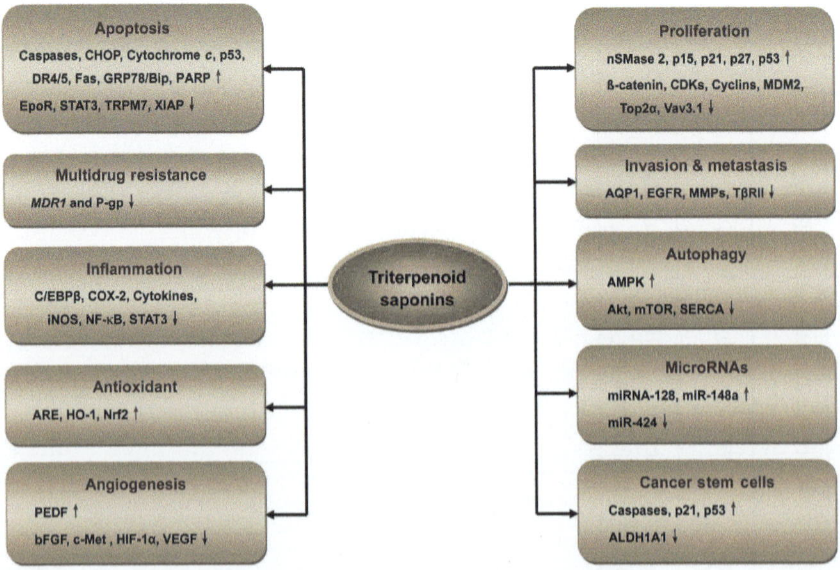

Figure 6.2 Potential molecular targets of triterpenoid saponins against cancer. ALDH1A1, aldehyde dehydrogenase 1A1; AMPK, AMP-activated protein kinase; AQP1, aquaporin 1; ARE, antioxidant response element; bFGF, basic fibroblast growth factor; C/EBPβ, CCAAT/enhancer binding protein β; CHOP, C/EBP homology protein; c-Met, c-mesenchymal-epithelial transition factor; COX-2, cyclooxygenase-2; CDKs, cyclin-dependent kinases; DR4/5, death receptor 4 and 5; EpoR, erythropoietin receptor; EGFR, epidermal growth factor receptor; GRP78/Bip, glucose-regulated protein 78/Bip; HIF-1α, hypoxia-inducible factor-1α; HO-1, heme oxygenase 1; iNOS, inducible nitric oxide synthase; *MDR1*, multidrug resistance gene 1; mTOR, mammalian target of rapamycin; MMPs, matrix metalloproteinases; NF-κB, nuclear factor-kappa B; Nrf2, nuclear factor (erythroid-derived 2)-like 2; nSMase 2, neutral sphingomyelinase 2; PEDF, pigment epithelium-derived factor; STAT3, signal transducer and activator of transcription-3; SERCA, sarcoplasmic/endoplasmic reticulum Ca^{2+} ATPase; TβRII, transforming growth factor-beta receptor type II; TRPM7, transient receptor potential melastatin 7; VEGF, vascular endothelial growth factor; XIAP, x-linked inhibitor of apoptosis proteins. (See the color plate.)

several types of ginsenosides [50,64,72,73], and sapogenins [80,82], which intervene with several diverse targets.

For example, astragaloside IV was observed to attenuate the colonogenic survival and anchorage-independent growth of HepG2 cells, likely a result of its effect on protein expression [24]. Specifically, astragaloside IV treatment resulted in decreased expression of the oncogene Vav3.1 implicating this protein in astragaloside IV's mechanism of action. β-Escin has been reported as a useful agent for colon cancer chemoprevention and treatment [17]. One study showed that administration of dietary 0.025% and

0.05% β-escin significantly suppressed azoxymethane-induced formation of aberrant colonic crypt foci. In cultured human HT-29 colon cancer cells, β-escin dose-dependently inhibited cellular proliferation by inducing expression of the cyclin-dependent kinase (CDK) inhibitor $p21^{waf1/cip1}$ in a p53-independent pathway and by increasing apoptosis through an unknown caspase-3-independent pathway. Echinoside A, a new marine-derived anticancer saponin, has been reported to show strong *in vitro* and *in vivo* anticancer properties in a panel of cancers [41]. In this study, Echinoside A displayed potent cytotoxicity in 24 human cell lines and two mouse cell lines tested, as well as total equipotent cytotoxicity between multidrug-resistant (MDR) cells and their corresponding parental cells (K562 vs. K562/A02, MCF-7 vs. MCF-7/ADR, KB vs. KB/VCR), suggesting that Echinoside A is a broad-spectrum anticancer agent without the potential of MDR induced by P-glycoprotein (P-gp) [41]. Moreover, Echinoside A also exhibited anticancer effects *in vivo*, as indicated by a 53.7% growth reduction in nude mice bearing human PC-3 prostate tumors, along with a 66.1% and 80% growth inhibition rate in murine H22 hepatoma and S180 sarcoma mouse models, respectively [41]. This study demonstrated that Echinoside A exerted potent *in vitro* and *in vivo* antitumor activities by targeting Top2a: Echinoside A interfered with its ability to bind DNA and impaired its ability to cleave and relegate DNA.

The antiproliferative effects of the ginsenosides Rh2 and Rg3 have been well-documented in cancers [64,72,73,75]. For example, Rh2 blocked mitosis at the G1 phase thus inhibiting the proliferation of HL-60 and U937 human leukemia cells. This was accompanied by a decrease in expression of several CDKs, an increase in expression of CDK inhibitors (CDKIs) such as $p21^{waf1/cip1}$ and $p27^{kip1}$, and enhanced binding of CDKIs to CDKs, resulting in reduced CDK activation [72]. Rh2 was also shown to induce leukemia cell differentiation. In this study, the authors showed that Rh2 regulation of TGF-β expression resulted in G1 arrest and differentiation of leukemia cells [72]. In another study, Rh2 was shown to significantly inhibit the growth of both MCF-7 and MDA-MB-231 breast cancer cells due to a cell cycle arrest at the G0/G1 phase. Authors showed this to occur through a $p15^{Ink4B}$ and $p27^{Kip1}$-dependent inhibition of G1-S specific Cdks/cyclin complexes [73]. The effect of Rh2 on cell cycle regulatory proteins was also observed in A549 lung cancer cells where it induced G1 growth arrest [75].

Rg3 has been shown to restrain colorectal cancer by downregulating the Wnt/ß-catenin signaling pathway. In this study, it dose-dependently inhibited cellular proliferation and colony formation of HCT116 human colorectal cancer cells *in vitro*, and also effectively inhibited tumor growth in nude mice bearing HCT116 xenografts [64]. Moreover, Rg3 decreased the nuclear content of ß-catenin, thus inhibiting ß-catenin/Tcf transcriptional activity. Interestingly, allelic deletion of ß-catenin in HCT116 cells rendered cancer cells more sensitive to Rg3-induced growth inhibition. These findings suggest that Rg3 is a potential agent for the clinical management of colorectal cancers through inhibition of Wnt/β-catenin signaling.

Protopanaxadiol and 20(*S*)-25-methoxyldammarane-3β,12β,20-triol (25-OCH_3-protopanaxadiol) are two sapogenins that have potent anticancer properties [80,82]. It was found that protopanaxadiol inhibited cellular proliferation through its modulation of cell cycle regulators and apoptotic proteins, leading to increased cell death in cultured K562 cells. Protopanaxadiol also reduced tumor growth in nude mice bearing K562 xenografts [80]. This study demonstrated that activation of neutral sphingomyelinase 2 was responsible for the anticancer activity of protopanaxadiol. Additionally, 25-OCH_3-protopanaxadiol was shown to exert potent inhibitory effects in both primary and metastatic breast cancers by inhibiting oncogene MDM2 expression at both transcriptional and post-translational levels [82].

3.2. Induction of Apoptosis and Autophagy

3.2.1 Apoptosis

Apoptosis, known as programmed cell death, plays a fundamental role in the balance of cell survival and cell death through two different pathways: the extrinsic death receptor-mediated pathway and the intrinsic mitochondria-mediated pathway. These apoptotic signaling pathways lead to the activation of caspases, resulting in the cleavage of different substrates and ultimately cell death [98]. A number of triterpenoid saponins have been reported to trigger cancer cell death by targeting signaling intermediates in both the intrinsic and extrinsic apoptotic pathways. Of note, the coactivation of both pathways has also been observed to contribute to the anticancer effect of several triterpenoid saponins including Avicin D, Platycodin D, Rg3, Rh2, and Tubeimoside-1.

Avicin D has been shown to induce apoptotic cell death of leukemia and lymphoma cells by activating the intrinsic caspase pathway through direct perturbation of mitochondria and downregulation of various prosurvival and antiapoptotic proteins [6,9]. Additionally, Avicin D has been shown to activate the Fas-caspase-8 pathway through the redistribution of Fas in lipid rafts [7]. Platycodin D has been shown as a potent apoptosis-inducing agent in a panel of human cancer cells [84,85,87–89]. The antitumor activity of Platycodin D has been exhibited by its induction of G2/M arrest and apoptosis in HepG2 human hepatoma cells [87]. In this study, authors observed a downregulation in protein expression of PI3K, P-Akt, and Bcl-2, and an upregulation in cleaved caspase-3, caspase-9, and PARP. This change in protein expression led to an inactivation of the PI3K/Akt signaling pathway, an increase in the Bax/Bcl-2 ratio, and ultimately activation of caspase-3, caspase-9, and PARP in Platycodin D-treated HepG2 cells. Moreover, a specific inhibitor for caspase-9 or caspase-3, but not for caspase-8 was able to significantly block Platycodin D-induced apoptosis in HepG2 cells, while concomitant Akt inhibition displayed a synergistic antiapoptotic effect with PD. Collectively, the antitumor activity of Platycodin D was attributable to its inhibition of PI3K/Akt-mediated intrinsic apoptosis pathway [87]. In breast cancer and gastric cancer cells, the coactivation of extrinsic and intrinsic pathways, together with activation of Akt or MAPK (ERK1/2, JNK, and p38) pathways, contributed to Platycodin D-induced apoptosis [84,88,89]. Interestingly, Platycodin D was also identified as a novel microtubule-disrupting agent in leukemia cells. Here, Platycodin D-induced apoptosis by interfering with cytoskeleton assembly along with several signaling pathways [85].

Rg3 has been shown to induce cellular apoptosis through the mitochondria-mediated caspase cascade inhuman and murine liver cancer cells [57,58]. Rg3 alone (3.0 mg/kg) or in combination with cyclophosphamide (20.0 mg/kg) significantly delayed tumor growth and prolongated animal survival in mice bearing Hep1-6 hepatoma xenografts by activating the intrinsic apoptosis pathway [58]. Downregulation of antiapoptotic molecules such as survivin, Pim-3, and p-Bad by Rg3 also contributed to its anticancer effects in SK-MES-1 human lung squamous cell carcinoma cells and PANC-1 pancreatic cancer cells [61,63]. TRAIL, a member of the TNF family of ligands, is regarded as a promising anticancer agent due to its selective induction of apoptosis in tumor cells without affecting normal cells. TRAIL mediates tumor cell apoptosis through the DR4 and DR5 death receptor-mediated extrinsic apoptosis pathway. Using this pathway, however, TRAIL resistance may develop due to the low expression of DR4 and DR5 in cancers in conjunction with a potential for increased expression of antiapoptotic molecules [99]. Noteworthy, 20(*S*)-ginsenoside Rg3 was observed to selectively promote TRAIL-induced apoptosis in various human liver cancer cells (Hep3B, HepG2, Huh-7, and SK-Hep1), but not in normal HL-7702 hepatocytes [67]. Moreover, chemosensitization of Rg3 (20 mg/kg) to TRAIL (3 mg/kg) occurred in nude mice bearing Huh-7 xenografts. This study indicated that Rg3 upregulation of DR5 through C/EBP homology protein (CHOP) led to the increase in sensitization of cells to TRAIL-induced apoptosis [67]. In another study, Kim *et al.* reported that treatment of human gastric adenocarcinoma AGS cells with Rg3 caused significant cytotoxicity and an increase in sub-G1, caspase-3 activity, and PARP cleavage [59]. Interestingly, blocking the transient receptor potential cation channel, subfamily M, member 7 (TRPM7) pharmacologically (La^{3+} or 2-APB) or with a siRNA directed against TRPM7 suppressed the *in vitro* anticancer effect of Rg3, while TRPM7 overexpression enhanced Rg3-induced cell apoptosis [59]. Taken together, these results suggest that Rg3 is a promising anticancer therapeutic agent, chemosensitizing cells to TRAIL, and potentially mediating its effects through TRPM7.

Among the six tested ginsenosides (Rc, Rd, Re, Rg3, Rh1, and Rh2), Rh2 displayed the highest cytotoxicity in HepG2 cells [71]. Rh2 exerted cytotoxicity via induction of apoptosis in cancer cells, and this could be attenuated by overexpressing Akt [74]. This study demonstrated that Rh2 induced internalization of lipid rafts and caveolae inactivated Akt in the plasma membrane. Inactivation of Akt led to a decrease in Bad phosphorylation with an increase in proapoptotic proteins Bim and Bax, thereby inducing mitochondria-mediated apoptosis. Rh2 has also been shown to induce apoptosis in human breast cancer cells *in vitro* and *in vivo* through mitochondria-mediated apoptosis [78]. In A549 human adenocarcinoma cells, Rh2 induced apoptosis via the extrinsic death receptor-mediated pathway, as indicated by upregulation of the TRAIL-RI (DR4) death receptor and subsequent activation of caspase-8 and -3 [75]. In addition, Rh2 was shown to induce Fas receptor aggregation in lipid rafts and subsequent caspase-8 activation in a FasL-independent manner in HeLa human cervical carcinoma cells [76]. Interestingly, the selective anticancer activity of Rh2 was observed more in HeLa cells than in any other cancer cell lines tested (HepG2 hepatoma, DU145 prostate carcinoma, HCT116 colon cancer) [71]. The authors found that both AMP-activated protein kinase (AMPK) and p38 MAPK acted as survival factors in Rh2-treated cancer cells, suggesting that inhibition of AMPK or p38 might increase the therapeutic potential of Rh2 [71].

Lastly, tubeimoside-1 has been shown as an efficient proapoptotic agent in various human cancer cells *in vitro* [25–30]. These studies showed that the broad therapeutic activity of Tubeimoside-1 is closely associated with activation of both the extrinsic and intrinsic caspase cascades and induction of endoplasmic reticulum stress through the upregulation of GRP78/Bip and CHOP.

3.2.2 Autophagy

Autophagy is defined as lysosomal-mediated degradation and recycling of intercellular components in order to maintain cellular homeostasis. Although autophagy has been well established as a cytoprotective mechanism in response to cellular stress such as starvation, accumulating evidence suggests that inadequate levels of autophagy can lead to caspase-independent autophagic cell death (also known as type II cell death) by excessive depletion of cellular organelles and essential proteins. Autophagy and apoptosis have a confounding relationship: protective autophagy results in the promotion of cell survival by inhibiting apoptosis; yet autophagy and apoptosis may also be cooperative in the induction of cell death. In the case of human cancer cells which are resistant to apoptotic therapies, the use of autophagic inducers capable of inducing type II cell death may present a novel therapeutic approach to overcome drug resistance [100].

In one study, Avicin D was reported to trigger autophagic cell death in AMPK-activated, apoptosis-resistant tumor cells [8]. In another study, Akebia saponin PA was shown to cause autophagic cell death in gastric cancer cells via PI3K/Akt/mTOR and AMPK/mTOR, which further activated p38/JNK MAPKs signal to promote caspase-3-dependent apoptosis [37]. Saikosaponin-D was newly identified as an autophagic inducer in several types of cancer cells by its direct inhibition of the sarcoplasmic/endoplasmic reticulum Ca^{2+}ATPase pump [32]. Moreover, Saikosaponin-D was also shown to be a potent cytotoxic agent in apoptosis-defective and apoptosis-resistant cells [32]. These studies suggest that Avicin D, Akebia saponin PA, and Saikosaponin-D may be potential therapeutic agents for apoptosis-resistant cancers.

In contrast, certain triterpenoids have been recently reported to induce protective autophagy, retarding apoptotic cell death, and thus limiting their therapeutic efficacy. Jujuboside B has been shown to induce apoptosis in AGS and HCT 116 gastric cancer cells and significantly suppress tumor growth in a nude mouse xenograft model. Authors showed that mice treated with Jujuboside had an increase in FasL and caspase-8 activation, resulting in activation of the p38/JNK-dependent extrinsic apoptosis pathway [97]. F2 was observed to induce apoptotic cell death in breast cancer stem cells

(CSCs) by activating the intrinsic caspase pathway [45]. However, both Jujuboside B and F2 induced protective autophagy, ultimately retarding apoptosis. Thus, the therapeutic potential of either of these drugs could be effectively improved with the combined use of an autophagy inhibitor [45,97]. These findings provide new insights into the rational use of triterpenoids as the potential anticancer agents.

The more complex crosstalk between autophagy and apoptosis has been described regarding the therapeutic properties of cucurbitacins *in vitro* [95]. While treatment with cucurbitacin-B or -I resulted in autophagic cell death through mitochondrial ROS-mediated ERK and JNK activation, pharmacological blockade of autophagy triggered a shift to apoptotic cell death in a caspase-dependent manner by inhibiting STAT3 signaling. Moreover, the cell-type differences and/or cellular levels of STAT3 seemed to modulate the balance between autophagic and apoptotic modes of cell death in cucurbitacin-treated cancer cells [95].

3.3. Attenuation of Invasion and Metastasis

Despite multidisciplinary treatment with surgery, chemotherapy, and radiation, human cancers with a high degree of local invasiveness and/or distant metastasis have poor clinical outcomes worldwide. Cancer invasion and metastasis are complex, multistep biochemical processes involving various physiopathological changes. With the identification of new molecular targets strongly linked to invasion and metastasis, the development of anti-invasive and antimetastatic agents is expected to provide an important therapeutic strategy for cancer. Noteworthy, diverse triterpenoid saponins such as Ardipusilloside I, Kalopanaxsaponin-A, Platycodin D, Rddeanin A, Rd, Rg3, Rh1, and Saikosaponin-D have exhibited significant anti-invasive and antimetastatic properties against a panel of human cancer cells *in vitro* and *in vivo* by intervening with signaling pathways or molecular mediators involved in invasion and metastasis [20,21,34,42,49,62,68,86,89].

Ginsenoside Rd, used at concentrations ranging from 10 to 100 μM, had no cytotoxic effect on the highly metastatic HepG2 liver cancer cells, but significantly inhibited their migration and invasion. Authors showed decreased expression of MMP-1,-2, and -7 through inactivation of MAPKs (ERK1/2, JNK, and p38) and transcription factor AP-1, along with reduced epithelial-to-mesenchymal transition (EMT), as indicated by increased E-cadherin expression and decreased N-cadherin expression [49]. Additionally, Rg3 inhibited cancer cell migration through p38 activation and subsequent downregulation of the aquaporin 1 water channel protein [62].

Use of triterpenoid saponins as anti-invasive and antimetastatic therapies has been widely evaluated in nude mice bearing several types of human cancer cells. For example, oral administration of Kalopanaxsaponin-A (1 or 5 mg/kg, three times a week for 25 days) dose-dependently inhibited orthotopic tongue tumor growth in nude mice bearing YD-10B oral squamous cell carcinoma cells and reduced the expression of several proteins involved in invasion such as MMP-9, HuR, and Rab1A [42]. Similarly, Platycodin D, administered at 5 mg/kg intraperitoneally for 20 days, greatly inhibited MDA-MB-231 metastatic breast cancer growth in nude mice [89]. In addition, while 15 μM Platycodin D had no cytotoxic effect, it significantly inhibited cell migration and adhesion, the expression of MMP-9 and EGFR, and MAPK activity in *in vitro*. Collectively, these results imply that the triterpenoids Kalopanaxsaponin-A, Platycodin D, Rd, and Rg3 may be potential anti-invasive and antimetastatic candidates for breast, hepatic, prostate, and oral cancers.

3.4. Inhibition of Angiogenesis

Angiogenesis, a process defined as formation of new blood vessels from preexisting structures, is a crucial step in cancer growth and metastasis as it supplies oxygen and nutrients to the growing tumor mass. The therapeutic properties of certain triterpenoid saponins such as Astragalus saponins, *Pulsatilla* saponin D (SB365), senegin III, and Rg3 have been reported as potential antiangiogenic agents associated with neovascular inhibition of endothelial cells in several cancers [22,54–56,90,91].

For example, Rg3 has been shown to reduce VEGF expression in various cancer cells and endothelial progenitor cells (EPCs), leading to an inhibition in angiogenesis and ultimately tumor growth [54–56]. *In vitro*, it has been reported that Rg3 inhibits the proliferation, migration, and tubular formation of EPCs. Oral administration (600 μg/kg/day) in Lewis lung carcinoma (LLC)-bearing mice resulted in a significant suppression of tumor growth and angiogenesis by inhibiting EPC mobilization from the bone marrow microenvironment to peripheral circulation [54]. The antitumor and antiangiogenic effects of Rg3 appeared to be related to the inhibition of VEGF-dependent p38/ERK signaling. Another study showed that combining Rg3 therapy (20 mg/kg/day) with low-dose gemcitabine exerted a synergistic therapeutic efficacy in LLC-bearing mice, which was attributable to the antiangiogenic effect of Rg3 through decreased VEGF expression [56].

SB365 has also been shown to exert a significant anticancer effect against human gastric cancer *in vitro* and *in vivo* [91]. At 0.1–20 μM, SB365 effectively inhibited the growth of gastric cancer cells MKN-45, MKN-28, and AGS, and this was attributable to its inhibition of proliferation along with its induction of apoptosis. Importantly, SB365 showed no cytotoxic effects in Hs677 normal gastric cells. Moreover, oral administration of SB365 (10 or 30 mg/kg/day) dose-dependently and significantly delayed tumor growth and also reduced tumor volumes in nude mice bearing MKN-45 gastric tumors. Notably, this reduced tumor growth was accompanied by a reduction in proliferation and angiogenesis along with an increase in apoptosis [91]. In addition, SB365 exerted potent antiangiogenic activity in hepatocyte growth factor-induced human umbilical vein endothelial cells (HUVEC). In this study, SB365 effectively inhibited endothelial tube formation *in vitro*, microvessel sprouting from the rat aortic ring *ex vivo*, and blood vessel formation using a Matrigel plug assay *in vivo*. This study also showed that SB365 could inhibit c-Met signaling in gastric cancer cells by docking at an allosteric site on c-Met, increasing the Bax/Bcl-2 ratio and activating caspase-3. The results from this study imply that inhibition of c-Met signaling, inhibition of cell growth and angiogenesis, and induction of apoptosis are responsible for the therapeutic effect of SB365 against gastric cancer.

Another type of triterpenoid saponin, the senegasaponins (e.g., senegin II, senegin III, senegin IV, senegasaponin a, and senegasaponin b), have been shown to exhibit selective antiproliferative activity against HUVECs with IC50 values in the range 0.6–6.2 μM with a selective index 7- to 100-fold greater than in several cancer cell lines (e.g., human epidermoid carcinoma KB3-1 cells, human leukemia K562 cells, and murine sarcoma S180 cells) [90]. Specifically, the antiangiogenic effect of senegin III verifiably inhibited VEGF-induced tubular formation of HUVECs *in vitro* and bFGF-induced neovascularization in the mouse Matrigel plug assay *in vivo* [90]. This study showed that senegin III induced the production of pigment epithelium-derived factor (PEDF), an endogenous inhibitor of angiogenesis, in HUVECs. Furthermore, senegin III (2.5 mg/kg) in combination with the anticancer drug cisplatin (1.0 mg/kg) synergistically improved tumor outcome in ddY mice bearing murine S180 sarcoma cells; it may be speculated that this synergy was likely associated with the antiangiogenic effect of senegin III.

3.5. Anti-inflammatory Effects

Epidemiological studies indicate that chronic inflammation is associated with increased cancer risk. Accumulating evidence suggests that the NF-κB and STAT3 pathways provide key molecular links between chronic inflammation and carcinogenesis. NF-κB and STAT3 are often constitutively activated in a variety of human cancer cells. This leads to the expression of transcription factor-regulating genes and subsequent proliferation, invasion, angiogenesis, and ultimately survival of cancer cells [101]. Therefore, anti-inflammatory intervention by way of targeting the NF-κB or STAT3 signaling pathways has the potential for prevention and treatment of inflammatory-associated cancers [102,103]. Several triterpenoid saponins, such as Avicin D, β-escin, Ds-echinoside A, ginsenoside Rg3, Saikosaponin-D, and Soyasaponin Bb, have been reported to inhibit cancer cell growth and progression through inactivation of inflammatory signaling [10,12,15,16,33,40,52,53,83].

β-Escin is a newly identified inhibitor of both NF-κB and STAT3 signaling pathways in human cancer cells [12,15,16]. Recent studies have shown that β-escin can inhibit constitutive and TNFα-inducible NF-κB activation in KBM-5 leukemia cells, as well as constitutive and interleukin-6-inducible STAT1 and STAT3 activation in HepG2, HUH-7, and PLC/PRF5 liver cancer cells and A549 lung cancer cells, thereby enhancing expression of proinflammatory gene products [12,15,16]. In addition, escin was shown to reduce the activation of p38 MAPK in A549 cells ultimately inhibiting their survival in a dose-dependent fashion. Interestingly, escin had no effect on the viability of human lens epithelial cells (HLECs) [12,15,16]. Lastly, in HepG2 cells, escin was shown to potentiate the apoptotic effect of doxorubicin and paclitaxel, indicating its promise in combination therapies [16].

Ginsenoside Rg3 was reported to exert significant therapeutic effects in human prostate cancer cells (LNCaP, PC3, and DU145) and colon cancer cells (SW620 and HCT116) by inactivating NF-κB signaling [52,53]. The combination treatment of Rg3 (50 μM) with docetaxel or other chemotherapeutics (cisplatin, doxorubicin, and orpaclitaxel) was found to be more effective than any of these agents alone. Combination drug therapy resulted in the inhibition of cancer cell growth and the induction of apoptosis along with G0/G1 cell cycle arrest through the inhibition of NF-κB and ultimately its target gene expression, such as those involved in apoptosis and cellular proliferation.

Ds-Echinoside A, a nonsulfated triterpene glycoside, exhibited significant cytotoxic activity with an IC50 of 2.65 μM in HepG2 human hepatocellular carcinoma cells [83]. Additionally, Ds-echinoside A dose-dependently inhibited the adhesion, migration, and invasion of HepG2 cells as well as the expression of NF-κB and MMP-9. These results imply that inhibition of NF-κB-dependent MMP-9 expression likely contributes to the antiproliferative and antimetastatic activity of Ds-echinoside A in hepatocellular carcinoma *in vitro* [83].

3.6. Antioxidative Effects

Oxidative stress-induced injury is known to be involved in the initiation, promotion, and progression of cancer. A number of studies have demonstrated that triterpenoids exhibit potent antioxidant activity by direct ROS scavenging and upregulation of antioxidant and/or detoxification enzymes through activating theNrf2-ARE pathway. Ginsenosides Rb1, Rg1, 20(*S*)-protopanaxatriol, or any low-dose combination thereof have been shown to synergistically enhance the total antioxidant activity of HepG2-C8 human liver hepatocellular carcinoma cells through upregulation of Nrf2 mRNA levels, increased ARE transcriptional activity, and subsequent induction of Nrf2-related genes products [47]. Rg3 was observed to enhance the chemosensitivity of CT-26 colon cancer cells to cisplatin and attenuate cisplatin-induced kidney and liver damage *in vitro* and *in vivo* by scavenging intracellular ROS [60].

In contrast, it is worth mentioning that the therapeutic properties of triterpenoid saponins can also contribute to prooxidant activity. Saxifragifolin D was observed to induce ROS accumulation and endoplasmic reticulum stress in MCF-7 and MDA-MB-231 breast cancer cells. However, this increase in prooxidant activity still led to the inhibition of cancer cell growth due to the delicate interplay between apoptosis and autophagy. Interestingly, pretreatment with the exogenous antioxidant *N*-acetyl-L-cysteine could suppress endoplasmic reticulum stress, apoptosis and autophagy induced by Saxifragifolin D, indicating the specificity of this effect [19].

3.7. Inhibition of Multidrug Resistance

Resistance to chemotherapy is a major challenge in clinical cancer treatment and is frequently associated with MDR. Extensive studies have shown that MDR in cancer cells involves complicated molecular mechanisms, among which overexpression of P-gp, encoded by the *MDR1* gene, acts as a drug efflux pump to extrude chemotherapeutic drugs out of MDR cancer cells. Inhibition of P-gp expression and/or transport function can restrain the P-gp-mediated MDR phenotype and improve the effectiveness of chemotherapy. Therefore, P-gp is widely regarded as an important target of reversal agents for MDR cancers.

Several triterpenoid saponins have been reported to increase the chemosensitivity of various kinds of cancer cells to conventional anticancer drugs, such as 5-fluorouracil [14], cisplatin [53,60,80], docetaxel [52,53], doxorubicin [53,80], gemcitabin [56], paclitaxel [16,66], and TRAIL [67]. The potential mechanisms of triterpenoid compounds against MDR were further explored in a panel of drug-resistant cancer cell lines *in vitro* and *in vivo*.

First, several groups have demonstrated a significant inhibition of MDR by several triterpenoid ginsenosides, such as protopanaxatriol, Rd, Rh2, and Rg3 [46,48,65,66,77,79]. Kim *et al.* reported the anti-MDR activity of Rg3 *in vitro* and *in vivo* [65]. In this study, Rg3 dose-dependently inhibited P-gp-mediated drug accumulation in KBV20C cells (an MDR human fibroblast carcinoma cell line), but not in drug-sensitive cells. This study showed that Rg3 reversed MDR to doxorubicin in drug-resistant cells by competing with the drug to bind P-gp, thereby blocking drug efflux. This MDR reversal significantly increased the life span of mice implanted with doxorubicin-resistant P388 murine leukemia cells [65]. Rg3 has also been reported to enhance the penetration of paclitaxel into Caco-2 monolayer cells by downregulating the expression of P-gp [66]. In another study, it was shown that Rh2 and 20(*S*)-Rh2 significantly chemosensitized MDR human breast cancer cells to paclitaxel or adriamycin by inhibition of P-gp function [77,79]. The chemosensitizing effect of protopanaxatriol has been shown in daunorubicin-resistant acute myelogenous leukemiacells (AML-2/D100). Here, protopanaxatriol was shown to increase the intracellular accumulation of drugs through direct interaction with P-gp at the azidopine site [46], while Rd reduced P-gp protein expression by upregulating its ubiquitin-dependent degradation [48]. Astragaloside II (0.08 mg/ml) strongly chemosensitized resistant Bel-7402/FU human hepatic cancer cells to 5-fluorouracil and inhibited P-gp transport function by inhibiting P-gp protein expression and *MDR1* gene expression [23].

3.8. Inhibition of CSCs

CSCs, with distinctive properties of self-renewal and the potential to proliferate extensively, have been identified in several common human cancers (e.g., hepatoma, breast, and lung cancer). They have been regarded as the root of oncogenesis, cancer relapse, and metastasis and are also involved in resistance to conventional cancer therapies. Therefore, the specific targeting and eradication of the CSC subpopulations is expected to provide an effective approach in the treatment of cancers. Oral administration of 500 ppm β-escin significantly suppressed lung tumor formation and delayed the progression of tobacco-induced adenocarcinoma formation in female A/J mice [18]. In another study, β-escin treatment in murine lung cancer cells and H460 human lung cancer cells was found to decrease ALDH-positive stem cells. Authors showed that β-escin induced expression of p21 in ALDH-positive and -negative lung cancer cells. These results suggest that the growth arrest of both cancer cells and CSCs is the mechanism of action of β-escin [18]. Interestingly, F2 exhibited similar cell death-inducing activity in MCF-7 human breast cancer cells and breast CSCs without inducing toxicity in CCD-25Lu normal human fibroblasts. However, conventional anticancer agents such as tamoxifen, 4-hydroxytamoxifen, and doxorubicin and phytochemicals such as quercetin, baicalein, tangeretin, and nobiletin had much weaker suppressive effects on breast CSC cell viability compared with MCF-7 cells [45]. Further mechanistic analysis showed that F2-induced apoptosis in breast CSCs by activating p53 and the downstream mitochondrial caspase apoptosis pathway. Of note, F2-induced intrinsic apoptosis was accompanied by protective autophagy and pharmacological inhibition of autophagy resulted in enhanced F2-induced cell death in breast CSCs [45].

3.9. Modulation of MicroRNAs

MicroRNAs (miRNAs) are noncoding small RNA molecules (18–25 nucleotides) implicated in a variety of biological processes and diseases. They negatively regulate gene expression by directly binding the 3′ untranslated regions of their target genes. Approximately 50% of all known miRNAs have been reported to be located in cancer-associated genomic regions, implying the potential association of miRNAs with tumor growth and progression [104]. Recent studies have identified an increasing number of miRNAs deregulated in several diverse types of human cancer cells including lung cancer, glioblastoma, and hepatoma. miRNAs therefore may be promising molecular targets for the chemoprevention and chemotherapy of cancer [105]. Wu *et al.* reported that Rh2 therapy significantly upregulated 14 miRNAs and significantly downregulated 12 miRNAs in cultured U251 human glioblastoma cells. Specifically, upregulation of brain-enriched miRNA-128 appeared to be responsible for Rh2-induced cytotoxicity, apoptosis, caspase-3 activation, and the transcriptional expression of E2F3a (a miR-128 target gene). This study indicated that miRNA-128 is the downstream effector of Rh2, mediating its dose-dependent inhibition of glioma cell growth [69]. Another study newly identified 44 upregulated miRNAs and 24 downregulated miRNAs in response to Rh2 treatment in A549 human lung cancer cells. Among the miRNAs affected, the expression levels of miR-148a and miR-424 were significantly up- and downregulated by 9.59- and 12.4-fold, respectively [70]. Importantly, these miRNAs have been predicted to target several genes related to angiogenesis, apoptosis, chromatin modification, cell proliferation, and cell differentiation, suggesting several potential downstream targets of Rh2.

4. STRUCTURE–ACTIVITY RELATIONSHIPS OF ANTICANCER ACTIVITIES OF TRITERPENOID SAPONINS

Thirty-six triterpenoid saponins isolated from the roots of *Pulsatilla chinensis* were tested for their cytotoxic effects against A549, SGC-7901, and HL-7702 cell lines. The results from this study indicated that a free carboxyl group at C-28 of aglycone is essential for cytotoxicity and the oleanane-type saponins have greater potency than the lupane-type saponins [106].

In another study, nine oleanane-type saponins and eight lupane-type saponins isolated from the roots of *Pulsatilla koreana* were examined for their *in vitro* cytotoxic effects against A-549, HCT15, SK-MEL-2, and SK-OV-3 human cancer cell lines. Among the tested compounds, the α-L-rhamnopyranosyl-(1→2)-α-L-arabinopyranosyl group at C-3 and the free carboxylic group at C-28 of the oleanane skeleton were found to be crucial for saponin cytotoxicity. Interestingly, the biochemical addition of a fourth sugar moiety at C-4‴ of the third glucopyranosyl moiety and the presence of a hydroxyl group at C-23 of aglycone enhanced saponin antitumor activity *in vivo* [107].

In one study, six *platycodon* saponins and two prosapogenins were tested for their antiproliferative effects in seven types of cancer cell lines. It was found that all six *platycodon* saponins were effective, and the presence of sugar residues and an *O*-acetyl group on the C-2 or C-3 position of rhamnose along with a methyl group at C-4 significantly increased the compounds' cytotoxicity and apoptotic induction [84]. In another study, twelve triterpenoid saponins isolated from the fruits of *Aesculus pavia* were tested *in vitro* for their cytotoxicity against 59 cell lines from nine different human cancers. The results from this study indicated that the presence of acyl groups at C-21 and C-22 are essential for cell killing [108].

Dammarane-type saponins are the primary compounds active in ginseng. These saponins possess four *trans*-ring rigid steroid skeletons with modified side chains at C-20 and sugar residues attached to the hydroxyl group of the aglycone. The relationship between the chemical structure of ginsenosides and their therapeutic activities has been systematically reviewed [5,109]. It was found that a decrease in the number of sugar moieties and an increase in the number of lipophilic groups on the side chain at C-20 enhanced ginsenoside therapeutic efficacy by increasing cell membrane permeability. Interestingly, the therapeutic potential of ginsenosides containing a sugar substitute at C-6 was shown to be weaker than those containing a sugar substitute at C-3 or C-20. On the other hand, the reduction of the double bond at C-24 or C-25 along with the addition of a hydroxyl group or methoxy group at C-25 improves the therapeutic potential of various ginsenosides. For example, 25-OCH_3-protopanaxadiol and 25-OH-protopanaxadiol are more potent cancer therapeutics than their counterparts 25-OH-protopanaxatriol, protopanaxadiol, Rh2, and Rg3. This is due to their enhanced reduction in proliferation, increased induction of apoptosis, enhanced cell cycle regulation and ultimate inhibition of tumor growth *in vivo*. In addition, the stereostructure of the C-20 hydroxyl may be influential on the therapeutic nature of ginsenosides. For example, 20(*S*)-Rg3 possesses greater antiproliferative effects than its 20(*R*) counterpart.

5. CLINICAL STUDIES

Although numerous laboratory-based studies have revealed the therapeutic properties of triterpenoid saponins in cancer cell lines and xenograft tumor models, few clinical trials have been conducted to observe their effects in human patients. The reasons may be partly attributed to the complex structures of triterpenoid saponins, their limited abundance in nature, their unstable properties, or difficulties in purification and chemical synthesis. The limited clinical studies performed to date are summarized below.

There have been two studies which examined the effectiveness of ginseng on survival and quality of life in cancer patients. One was a randomized placebo-controlled study conducted in Korea among 53 patients with diverse cancers [110]. The other was a cohort study completed in China among 1455 breast cancer patients [111]. Both studies found that ginseng root and its by-products had positive effects in the physical, psychological, and social well-being domains. The Chinese study also found that regular users had a significantly reduced risk of death.

Another three studies reported that both Asian and American ginseng had positive effects on cancer-related fatigue. One study involving 232 patients with chronic functional fatigue concluded that patients given Asian ginseng had significantly improved fatigue symptom scores compared to patients given a placebo [112]. Another study which administered Asian ginseng in a small group of chemo-naive cancer patients (sample size $N=20$) reported that ginseng significantly improved total and average fatigue levels [113]. Lastly, a randomized Phase I clinical trial revealed that high doses of ginseng (1000 and 2000 mg/day) alleviated cancer-related fatigue significantly greater than placebo [113].

However, there also have been some conflicting results. For example, while a large placebo-controlled study and a small cohort study both suggested that ginseng use is associated with a 60–70% reduction in gastric cancer risk in Korean populations, another separate large cohort study, which attempted to duplicate these results among healthy Chinese women, found no association between ginseng intake and gastric cancer risk [114].

In general though, a fairly consistent protective role of ginseng has been found when assessing cancer risk and cancer mortality. For example, the results from a prospective cohort study which enrolled 4634 people over 40 years old suggested that ginseng had a nonorgan-specific preventative effect against cancer [115]. A similar conclusion was drawn through a randomized clinical trial in which 643 chronic atrophic gastritis patients were orally administered red ginseng extract weekly for 3 years and followed up for 8 years. The ensuing development of various cancers was compared to a placebo-controlled group. Results from this study indicated that administration of red ginseng extract for 3 years significantly decreased the incidence of nonorgan-specific human cancers in males [116].

6. SUMMARY AND PERSPECTIVES

Great progress has been made in understanding the therapeutic efficacy of triterpenoid saponins in the treatment of cancer. Increasing evidence demonstrates that diverse triterpenoid saponins exert potent anticancer properties by targeting multiple signaling molecules implicated in the carcinogenic process, as summarized in Table 6.1 and Fig. 6.2. Of note, several triterpenoid saponins have been shown to be selectively cytotoxic against cancer cells *in vitro* and *in vivo*, leaving nonneoplastic tissues unharmed. Indeed, administration of triterpenoid saponins in various animal models has demonstrated the high safety and nontoxic profile of these compounds. In this review, we have described the therapeutic effects of triterpenoid saponins for primary and metastatic cancer and also for cancers previously found drug-resistant. Additionally, the diversity in chemical structure of triterpenoid saponins raises the fundamental requirements for the elucidation of their structure–function relationship.

Taken together, triterpenoid saponins may prove to be the attractive candidates for the development of the ideal anticancer agents with prominent efficacy and desirable drugability. Future studies examining the overall therapeutic effectiveness in more extensive and well-controlled clinical trials are therefore necessary to fully understand the benefits of triterpenoid saponins in cancer therapy.

REFERENCES

[1] J.M. Augustin, et al., Molecular activities, biosynthesis and evolution of triterpenoid saponins, Phytochemistry 72 (6) (2011) 435–457.

[2] R. Ali, et al., New anticancer agents: Recent developments in tumor therapy, Anticancer Res. 32 (7) (2012) 2999–3005.

[3] S. Man, et al., Chemical study and medical application of saponins as anti-cancer agents, Fitoterapia 81 (7) (2010) 703–714.

[4] S.G. Sparg, M.E. Light, J. van Staden, Biological activities and distribution of plant saponins, J. Ethnopharmacol. 94 (2–3) (2004) 219–243.

[5] L.W. Qi, C.Z. Wang, C.S. Yuan, American ginseng: Potential structure-function relationship in cancer chemoprevention, Biochem. Pharmacol. 80 (7) (2010) 947–954.

[6] V. Haridas, et al., Avicins: Triterpenoid saponins from *Acacia victoriae* (Bentham) induce apoptosis by mitochondrial perturbation, Proc. Natl. Acad. Sci. U.S.A. 98 (10) (2001) 5821–5826.

[7] Z.X. Xu, et al., Avicin D, a plant triterpenoid, induces cell apoptosis by recruitment of Fas and downstream signaling molecules into lipid rafts, PLoS One 4 (12) (2009) e8532.

[8] Z.X. Xu, et al., A plant triterpenoid, avicin D, induces autophagy by activation of AMP-activated protein kinase, Cell Death Differ. 14 (11) (2007) 1948–1957.

[9] A. Gaikwad, et al., Triterpenoid electrophiles (Avicins) suppress heat shock protein-7 and X-linked inhibitor of apoptosis proteins in malignant cells by activation of ubiquitin machinery: Implications for proapoptotic activity, Clin. Cancer Res. 11 (5) (2005) 1953–1962.

[10] V. Haridas, et al., Avicin D: A protein reactive plant isoprenoid dephosphorylates Stat 3 by regulating both kinase and phosphatase activities, PLoS One 4 (5) (2009) e5578.

[11] M. Fukumura, et al., Achyranthoside H methyl ester, a novel oleanolic acid saponin derivative from *Achyranthes fauriei* roots, induces apoptosis in human breast cancer MCF-7 and MDA-MB-453 cells via a caspase activation pathway, J. Nat. Med. 63 (2) (2009) 181–188.

[12] K.B. Harikumar, et al., Escin, a pentacyclic triterpene, chemosensitizes human tumor cells through inhibition of nuclear factor-κB signaling pathway, Mol. Pharmacol. 77 (5) (2010) 818–827.
[13] D.Y. Shen, et al., Apoptosis of human cholangiocarcinoma cell lines induced by β-escin through mitochondrial caspase-dependent pathway, Phytother. Res. 25 (10) (2011) 1519–1526.
[14] Z.J. Ming, et al., Synergistic effects of β-aescin and 5-fluorouracil in human hepatocellular carcinoma SMMC-7721 cells, Phytomedicine 17 (8–9) (2010) 575–580.
[15] D.B. Ji, et al., β-Escin sodium inhibits inducible nitric oxide synthase expression via downregulation of the JAK/STAT pathway in A549 cells, Mol. Carcinog. 50 (12) (2011) 945–960.
[16] S.M. Tan, et al., Identification of β-escin as a novel inhibitor of signal transducer and activator of transcription 3/Janus-activated Kinase 2 signaling pathway that suppresses proliferation and induces apoptosis in human hepatocellular carcinoma cells, J. Pharmacol. Exp. Ther. 334 (1) (2010) 285–293.
[17] J.M. Patlolla, et al., β-Escin inhibits colonic aberrant crypt foci formation in rats and regulates the cell cycle growth by inducing $p21^{waf1/cip1}$ in colon cancer cells, Mol. Cancer Ther. 5 (6) (2006) 1459–1466.
[18] J.M. Patlolla, et al., β-Escin inhibits NNK-induced lung adenocarcinoma and ALDH1A1 and RhoA/Rock expression in A/J mice and growth of H460 human lung cancer cells, Cancer Prev. Res. (Phila) 6 (10) (2013) 1140–1149.
[19] J.M. Shi, et al., Saxifragifolin D induces the interplay between apoptosis and autophagy in breast cancer cells through ROS-dependent endoplasmic reticulum stress, Biochem. Pharmacol. 85 (7) (2013) 913–926.
[20] G. Xue, et al., Raddeanin A induces human gastric cancer cells apoptosis and inhibits their invasion in vitro, Biochem. Biophys. Res. Commun. 439 (2) (2013) 196–202.
[21] L. Lou, et al., Ardipusilloside inhibits survival, invasion and metastasis of human hepatocellular carcinoma cells, Phytomedicine 19 (7) (2012) 603–608.
[22] K.K. Auyeung, et al., Astragalus saponins modulate cell invasiveness and angiogenesis in human gastric adenocarcinoma cells, J. Ethnopharmacol. 141 (2) (2012) 635–641.
[23] C. Huang, et al., Reversal of P-glycoprotein-mediated multidrug resistance of human hepatic cancer cells by Astragaloside II, J. Pharm. Pharmacol. 64 (12) (2012) 1741–1750.
[24] H. Qi, et al., Proteomic characterization of the cellular response to chemopreventive triterpenoid astragaloside IV in human hepatocellular carcinoma cell line HepG2, Int. J. Oncol. 36 (3) (2010) 725–735.
[25] X.Y. Weng, R.D. Ma, L.J. Yu, Apoptosis of human nasopharyngeal carcinoma CNE-2Z cells induced by tubeimoside I, Ai Zheng 22 (8) (2003) 806–811.
[26] Y. Yin, et al., NF-κB, JNK and p53 pathways are involved in tubeimoside-1-induced apoptosis in HepG2 cells with oxidative stress and G2/M cell cycle arrest, Food Chem. Toxicol. 49 (12) (2011) 3046–3054.
[27] P. Huang, et al., Cytotoxicity of tubeimoside I in human choriocarcinoma JEG-3 cells by induction of cytochrome c release and apoptosis via the mitochondrial-related signaling pathway, Int. J. Mol. Med. 28 (4) (2011) 579–587.
[28] Y. Wang, et al., Natural plant extract tubeimoside I promotes apoptosis-mediated cell death in cultured human hepatoma (HepG2) cells, Biol. Pharm. Bull. 34 (6) (2011) 831–838.
[29] Y. Zhang, X. Xu, P. He, Tubeimoside-1 inhibits proliferation and induces apoptosis by increasing the Bax to Bcl-2 ratio and decreasing COX-2 expression in lung cancer A549 cells, Mol. Med. Rep. 4 (1) (2011) 25–29.
[30] W.J. Chen, et al., Tubeimoside-1 induces G2/M phase arrest and apoptosis in SKOV-3 cells through increase of intracellular Ca^{2+} and caspase-dependent signaling pathways, Int. J. Oncol. 40 (2) (2012) 535–543.

[31] B.M. Kim, S.H. Hong, Sequential caspase-2 and caspase-8 activation is essential for saikosaponin a-induced apoptosis of human colon carcinoma cell lines, Apoptosis 16 (2) (2011) 184–197.
[32] V.K. Wong, et al., Saikosaponin-d, a novel SERCA inhibitor, induces autophagic cell death in apoptosis-defective cells, Cell Death Dis. 4 (2013) e720.
[33] X.L. Lu, et al., Chemopreventive effect of saikosaponin-d on diethylinitrosamine-induced hepatocarcinogenesis: Involvement of CCAAT/enhancer binding protein β and cyclooxygenase-2, Mol. Med. Rep. 5 (3) (2012) 637–644.
[34] X. Jia, et al., Effects of saikosaponin-D on syndecan-2, matrix metalloproteinases and tissue inhibitor of metalloproteinases-2 in rats with hepatocellular carcinoma, J. Tradit. Chin. Med. 32 (3) (2012) 415–422.
[35] L.Y. Guo, et al., Cimiside E arrests cell cycle and induces cell apoptosis in gastric cancer cells, Arch. Pharm. Res. 32 (10) (2009) 1385–1392.
[36] S. Mallick, et al., Corchorusin-D directed apoptosis of K562 cells occurs through activation of mitochondrial and death receptor pathways and suppression of AKT/PKB pathway, Cell. Physiol. Biochem. 30 (4) (2012) 915–926.
[37] M.Y. Xu, et al., Akebia saponin PA induces autophagic and apoptotic cell death in AGS human gastric cancer cells, Food Chem. Toxicol. 59 (2013) 703–708.
[38] S. Mo, et al., Phaseoloideside E, a novel natural triterpenoid saponin identified from Entada phaseoloides, induces apoptosis in EC-109 esophageal cancer cells through reactive oxygen species generation, J. Pharmacol. Sci. 122 (3) (2013) 163–175.
[39] N.X. Nhiem, et al., Cytotoxic oleane-type triterpene saponins from *Glochidion eriocarpum*, Arch. Pharm. Res. 35 (1) (2012) 19–26.
[40] Y. Pei, et al., Apoptosis effects on human esophageal cancer cells by soyasaponin Bb and its mechanism, Wei Sheng Yan Jiu 39 (4) (2010) 444–446.
[41] M. Li, et al., Echinoside A, a new marine-derived anticancer saponin, targets topoisomerase2α by unique interference with its DNA binding and catalytic cycle, Ann. Oncol. 21 (3) (2010) 597–607.
[42] Y.S. Hwang, K.K. Park, W.Y. Chung, Kalopanaxsaponin A inhibits the invasion of human oral squamous cell carcinoma by reducing metalloproteinase-9 mRNA stability and protein trafficking, Biol. Pharm. Bull. 35 (3) (2012) 289–300.
[43] F. Guan, et al., Apoptosis and membrane permeabilisation induced by macranthoside B on HL-60 cells, Nat. Prod. Res. 25 (4) (2011) 332–340.
[44] P. Koczurkiewicz, et al., Triterpene saponosides from *Lysimachia ciliate* differentially attenuate invasive potential of prostate cancer cells, Chem. Biol. Interact. 206 (1) (2013) 6–17.
[45] T.T. Mai, et al., Ginsenoside F2 induces apoptosis accompanied by protective autophagy in breast cancer stem cells, Cancer Lett. 321 (2) (2012) 144–153.
[46] C.H. Choi, G. Kang, Y.D. Min, Reversal of P-glycoprotein-mediated multidrug resistance by protopanaxatriol ginsenosides from Korean red ginseng, Planta Med. 69 (3) (2003) 235–240.
[47] C.L. Saw, et al., Pharmacodynamics of ginsenosides: Antioxidant activities, activation of Nrf2, and potential synergistic effects of combinations, Chem. Res. Toxicol. 25 (8) (2012) 1574–1580.
[48] Y.R. Pokharel, et al., Increased ubiquitination of multidrug resistance1 by ginsenoside Rd, Nutr. Cancer 62 (2) (2010) 252–259.
[49] J.H. Yoon, et al., Anti-metastatic effects of ginsenoside Rd via inactivation of MAPK signaling and induction of focal adhesion formation, Phytomedicine 19 (3–4) (2012) 284–292.
[50] S. Cai, et al., Experimental study on human leukemia cell line K562 senescence induced by ginsenoside Rg1, Zhongguo Zhong Yao Za Zhi 37 (16) (2012) 2424–2428.

[51] J. Li, et al., Ginsenoside Rg1 induces apoptosis through inhibition of the EpoR-mediated JAK2/STAT5 signalling pathway in the TF-1/Epo human leukemia cell line, Asian Pac. J. Cancer Prev. 15 (6) (2014) 2453–2459.
[52] S.M. Kim, et al., Inhibition of NF-κB by ginsenoside Rg3 enhances the susceptibility of colon cancer cells to docetaxel, Arch. Pharm. Res. 32 (5) (2009) 755–765.
[53] S.M. Kim, et al., Combination of ginsenoside Rg3 with docetaxel enhances the susceptibility of prostate cancer cells via inhibition of NF-κB, Eur. J. Pharmacol. 631 (1–3) (2010) 1–9.
[54] J.W. Kim, et al., Ginsenoside Rg3 attenuates tumor angiogenesis via inhibiting bioactivities of endothelial progenitor cells, Cancer Biol. Ther. 13 (7) (2012) 504–515.
[55] Q.J. Chen, M.Z. Zhang, L.X. Wang, Gensenoside Rg3 inhibits hypoxia-induced VEGF expression in human cancer cells, Cell. Physiol. Biochem. 26 (6) (2010) 849–858.
[56] T.G. Liu, et al., Inhibitory effect of ginsenoside Rg3 combined with gemcitabine on angiogenesis and growth of lung cancer in mice, BMC Cancer 9 (2009) 250.
[57] C. Zhang, et al., Antitumor effects of ginsenoside Rg3 on human hepatocellular carcinoma cells, Mol. Med. Rep. 5 (5) (2012) 1295–1298.
[58] J.W. Jiang, et al., Ginsenoside Rg3 inhibit hepatocellular carcinoma growth via intrinsic apoptotic pathway, World J. Gastroenterol. 17 (31) (2011) 3605–3613.
[59] B.J. Kim, et al., Transient receptor potential melastatin 7 channels are involved in ginsenoside Rg3-induced apoptosis in gastric cancer cells, Basic Clin. Pharmacol. Toxicol. 109 (4) (2011) 233–239.
[60] C.K. Lee, et al., Ginsenoside Rg3 enhances the chemosensitivity of tumors to cisplatin by reducing the basal level of nuclear factor erythroid 2-related factor 2-mediated heme oxygenase-1/NAD(P)H quinone oxidoreductase-1 and prevents normal tissue damage by scavenging cisplatin-induced intracellular reactive oxygen species, Food Chem. Toxicol. 50 (7) (2012) 2565–2574.
[61] J. Jian, Z.F. Hu, Y. Huang, Effect of ginsenoside Rg3 on Pim-3 and Bad proteins in human pancreatic cancer cell line PANC-1, Ai Zheng 28 (5) (2009) 461–465.
[62] X.Y. Pan, et al., Ginsenoside Rg3 attenuates cell migration via inhibition of aquaporin 1 expression in PC-3 M prostate cancer cells, Eur. J. Pharmacol. 683 (1–3) (2012) 27–34.
[63] X. Wang, et al., Ginsenoside Rg3 induces apoptosis of human lung squamous cell carcinoma SK-MES-1 cell line, Nan Fang Yi Ke Da Xue Xue Bao 29 (9) (2009) 1823–1826.
[64] B.C. He, et al., Ginsenoside Rg3 inhibits colorectal tumor growth through the down-regulation of Wnt/ß-catenin signaling, Int. J. Oncol. 38 (2) (2011) 437–445.
[65] S.W. Kim, et al., Reversal of P-glycoprotein-mediated multidrug resistance by ginsenoside Rg3, Biochem. Pharmacol. 65 (1) (2003) 75–82.
[66] L.Q. Yang, et al., Enhanced oral bioavailability and anti-tumour effect of paclitaxel by 20(S)-ginsenoside Rg3 in vivo, Biopharm. Drug Dispos. 33 (8) (2012) 425–436.
[67] J.Y. Lee, et al., Sensitization of TRAIL-induced cell death by 20(S)-ginsenoside Rg3 via CHOP-mediated DR5 upregulation in human hepatocellular carcinoma cells, Mol. Cancer Ther. 12 (3) (2013) 274–285.
[68] J.H. Yoon, Y.J. Choi, S.G. Lee, Ginsenoside Rh1 suppresses matrix metalloproteinase-1 expression through inhibition of activator protein-1 and mitogen-activated protein kinase signaling pathway in human hepatocellular carcinoma cells, Eur. J. Pharmacol. 679 (1–3) (2012) 24–33.
[69] N. Wu, et al., Ginsenoside Rh2 inhibits glioma cell proliferation by targeting microRNA-128, Acta Pharmacol. Sin. 32 (3) (2011) 345–353.
[70] I.S. An, et al., Ginsenoside Rh2 mediates changes in the microRNA expression profile of human non-small cell lung cancer A549 cells, Oncol. Rep. 29 (2) (2013) 523–528.

[71] M.J. Kim, et al., AMP-activated protein kinase determines apoptotic sensitivity of cancer cells to ginsenoside-Rh2, J. Ginseng Res. 38 (1) (2014) 16–21.
[72] K.S. Chung, et al., Ginsenoside Rh2 induces cell cycle arrest and differentiation in human leukemia cells by upregulating TGF-β expression, Carcinogenesis 34 (2) (2013) 331–340.
[73] S. Choi, T.W. Kim, S.V. Singh, Ginsenoside Rh2-mediated G1 phase cell cycle arrest in human breast cancer cells is caused by $p15^{Ink4B}$ and $p27^{Kip1}$-dependent inhibition of cyclin-dependent kinases, Pharm. Res. 26 (10) (2009) 2280–2288.
[74] E.K. Park, et al., Induction of apoptosis by the ginsenoside Rh2 by internalization of lipid rafts and caveolae and inactivation of Akt, Br. J. Pharmacol. 160 (5) (2010) 1212–1223.
[75] C.C. Cheng, et al., Molecular mechanisms of ginsenoside Rh2-mediated G1 growth arrest and apoptosis in human lung adenocarcinoma A549 cells, Cancer Chemother. Pharmacol. 55 (6) (2005) 531–540.
[76] J.S. Yi, et al., Ginsenoside Rh2 induces ligand-independent Fas activation via lipid raft disruption, Biochem. Biophys. Res. Commun. 385 (2) (2009) 154–159.
[77] W.W. Jia, et al., Rh2, a compound extracted from ginseng, hypersensitizes multidrug-resistant tumor cells to chemotherapy, Can. J. Physiol. Pharmacol. 82 (7) (2004) 431–437.
[78] S. Choi, J.Y. Oh, S.J. Kim, Ginsenoside Rh2 induces Bcl-2 family proteins-mediated apoptosis in vitro and in xenografts in vivo models, J. Cell. Biochem. 112 (1) (2011) 330–340.
[79] J. Zhang, et al., Cellular pharmacokinetic mechanisms of adriamycin resistance and its modulation by 20(S)-ginsenoside Rh2 in MCF-7/Adr cells, Br. J. Pharmacol. 165 (1) (2012) 120–134.
[80] B. Park, et al., Neutral sphingomyelinase 2 modulates cytotoxic effects of protopanaxadiol on different human cancer cells, BMC Complement. Altern. Med. 13 (2013) 194.
[81] L.H. Zhang, et al., AD-1, a novel ginsenoside derivative, shows anti-lung cancer activity via activation of p38 MAPK pathway and generation of reactive oxygen species, Biochim. Biophys. Acta 1830 (8) (2013) 4148–4159.
[82] W. Wang, et al., Natural product ginsenoside 25-OCH_3-PPD inhibits breast cancer growth and metastasis through downregulating MDM2, PLoS One 7 (7) (2012) e41586.
[83] Q. Zhao, et al., Ds-echinoside A, a new triterpene glycoside derived from sea cucumber, exhibits antimetastatic activity via the inhibition of NF-κB-dependent MMP-9 and VEGF expressions, J. Zhejiang Univ. Sci. B 12 (7) (2011) 534–544.
[84] J. Chun, I.J. Ha, Y.S. Kim, Antiproliferative and apoptotic activities of triterpenoid saponins from the roots of *Platycodon grandiflorum* and their structure-activity relationships, Planta Med. 79 (8) (2013) 639–645.
[85] M.O. Kim, et al., Platycodin D induces mitotic arrestin vitro, leading to endoreduplication, inhibition of proliferation and apoptosis in leukemia cells, Int. J. Cancer 122 (12) (2008) 2674–2681.
[86] Y. Ye, et al., Combination treatment with platycodin D and osthole inhibits cell proliferation and invasion in mammary carcinoma cell lines, Environ. Toxicol. Pharmacol. 36 (1) (2013) 115–124.
[87] H. Qin, et al., Platycodin D, a triterpenoid saponin from *Platycodon grandiflorum*, induces G2/M arrest and apoptosis in human hepatoma HepG2 cells by modulating the PI3K/Akt pathway, Tumour Biol. 35 (2) (2014) 1267–1274.
[88] J.S. Yu, A.K. Kim, Platycodin D induces apoptosis in MCF-7 human breast cancer cells, J. Med. Food 13 (2) (2010) 298–305.

[89] J. Chun, Y.S. Kim, Platycodin D inhibits migration, invasion, and growth of MDA-MB-231 human breast cancer cells via suppression of EGFR-mediated Akt and MAPK pathways, Chem. Biol. Interact. 205 (3) (2013) 212–221.
[90] M. Arai, et al., Anti-angiogenic effect of triterpenoidal saponins from *Polygala senega*, J. Nat. Med. 65 (1) (2011) 149–156.
[91] S.W. Hong, et al., SB365, Pulsatillasaponin D, targets c-Met and exerts antiangiogenic and antitumor activities, Carcinogenesis 34 (9) (2013) 2156–2169.
[92] M.K. Son, et al., SB365, Pulsatillasaponin D suppresses the proliferation of human colon cancer cells and induces apoptosis by modulating the AKT/mTOR signalling pathway, Food Chem. 136 (1) (2013) 26–33.
[93] F. Guan, et al., Bigelovii A induces apoptosis of HL-60 human acute promyelocytic leukaemia cells, Mol. Med. Rep. 7 (5) (2013) 1585–1590.
[94] X. Zhu, et al., Ziyuglycoside II inhibits the growth of human breast carcinoma MDA-MB-435 cells via cell cycle arrest and induction of apoptosis through the mitochondria dependent pathway, Int. J. Mol. Sci. 14 (9) (2013) 18041–18055.
[95] T. Zhang, et al., Cucurbitacin induces autophagy through mitochondrial ROS production which counteracts to limit caspase-dependent apoptosis, Autophagy 8 (4) (2012) 559–576.
[96] G. Ishdorj, J.B. Johnston, S.B. Gibson, Inhibition of constitutive activation of STAT3 by curcurbitacin-I (JSI-124) sensitized human B-leukemia cells to apoptosis, Mol. Cancer Ther. 9 (12) (2010) 3302–3314.
[97] M.Y. Xu, et al., Antitumor activity of Jujuboside B and the underlying mechanism via induction of apoptosis and autophagy, J. Nat. Prod. 77 (2) (2014) 370–376.
[98] S. Fulda, K.M. Debatin, Extrinsic versus intrinsic apoptosis pathways in anticancer chemotherapy, Oncogene 25 (34) (2006) 4798–4811.
[99] L.Y. Dimberg, et al., On the TRAIL to successful cancer therapy? Predicting and counteracting resistance against TRAIL-based therapeutics, Oncogene 32 (11) (2013) 1341–1350.
[100] D.A. Gewirtz, Cytoprotective and nonprotective autophagy in cancer therapy, Autophagy 9 (9) (2013) 1263–1265.
[101] A.C. Bharti, et al., Nuclear factor-κB and STAT3 are constitutively active in $CD138^+$ cells derived from multiple myeloma patients, and suppression of these transcription factors leads to apoptosis, Blood 103 (8) (2004) 3175–3184.
[102] L. Klampfer, Signal transducers and activators of transcription (STATs): Novel targets of chemopreventive and chemotherapeutic drugs, Curr. Cancer Drug Targets 6 (2) (2006) 107–121.
[103] H.J. Kim, N. Hawke, A.S. Baldwin, NF-κB and IKK as therapeutic targets in cancer, Cell Death Differ. 13 (5) (2006) 738–747.
[104] L. Zhang, et al., MicroRNAs exhibit high frequency genomic alterations in human cancer, Proc. Natl. Acad. Sci. U.S.A. 103 (24) (2006) 9136–9141.
[105] A. Izzotti, et al., MicroRNAs as targets for dietary and pharmacological inhibitors of mutagenesis and carcinogenesis, Mutat. Res. 751 (2) (2012) 287–303.
[106] K. Xu, et al., Cytotoxic activity of *Pulsatilla chinensis* saponins and their structure-activity relationship, J. Asian Nat. Prod. Res. 15 (6) (2013) 680–686.
[107] S.C. Bang, et al., Antitumor activity of *Pulsatilla koreana* saponins and their structure-activity relationship, Chem. Pharm. Bull. 53 (11) (2005) 1451–1454.
[108] Z. Zhang, S. Li, Cytotoxic triterpenoid saponins from the fruits of *Aesculus pavia* L, Phytochemistry 68 (15) (2007) 2075–2086.
[109] S.A. Nag, et al., Ginsenosides as anticancer agents: In vitro and in vivo activities, structure-activity relationships, and molecular mechanisms of action, Front. Pharmacol. 3 (2012) 25.

[110] J.H. Kim, C.Y. Park, S.J. Lee, Effects of Sun Ginseng on subjective quality of life in cancer patients: A double-blind, placebo-controlled pilot trial, J. Clin. Pharm. Ther. 31 (4) (2006) 331–334.

[111] Y. Cui, et al., Association of ginseng use with survival and quality of life among breast cancer patients, Am. J. Epidemiol. 163 (7) (2006) 645–653.

[112] A. Caso Marasco, et al., Double-blind study of a multivitamin complex supplemented with ginseng extract, Drugs Exp. Clin. Res. 22 (6) (1996) 323–329.

[113] D.L. Barton, et al., Pilot study of *Panax quinquefolius* (American ginseng) to improve cancer-related fatigue: A randomized, double-blind, dose-finding evaluation: NCCTG trial N03CA, Support. Care Cancer 18 (2) (2010) 179–187.

[114] F. Kamangar, et al., Ginseng intake and gastric cancer risk in the Shanghai women's health study cohort, Cancer Epidemiol. Biomarkers Prev. 16 (3) (2007) 629–630.

[115] T.K. Yun, S.Y. Choi, Non-organ specific cancer prevention of ginseng: A prospective study in Korea, Int. J. Epidemiol. 27 (3) (1998) 359–364.

[116] T.K. Yun, et al., Non-organ-specific preventive effect of long-term administration of Korean red ginseng extract on incidence of human cancers, J. Med. Food 13 (3) (2010) 489–494.

CHAPTER SEVEN

Neem Limonoids as Anticancer Agents: Modulation of Cancer Hallmarks and Oncogenic Signaling

Siddavaram Nagini[1]
Faculty of Science, Department of Biochemistry and Biotechnology, Annamalai University, Annamalainagar, Tamil Nadu, India
[1]Corresponding author: e-mail address: s_nagini@yahoo.com; snlabau@gmail.com

Contents

Abstract

Neem (*Azadirachta indica A. Juss*) is one of the most versatile medicinal plants, widely distributed in the Indian subcontinent. Neem is a rich source of limonoids that are endowed with potent medicinal properties predominantly antioxidant, anti-inflammatory, and anticancer activities. Azadirachtin, gedunin, and nimbolide are more extensively investigated relative to other neem limonoids. Accumulating evidence indicates that the anticancer effects of neem limonoids are mediated through the inhibition of hallmark capabilities of cancer such as cell proliferation, apoptosis evasion, inflammation,

The Enzymes, Volume 36
ISSN 1874-6047
http://dx.doi.org/10.1016/B978-0-12-802215-3.00007-0

invasion, and angiogenesis. The neem limonoids have been demonstrated to target oncogenic signaling kinases and transcription factors chiefly, NF-κB, Wnt/β-catenin, PI3K/Akt, MAPK, and JAK/STAT signaling pathways. Neem limonoids that target multiple pathways that are aberrant in cancer are ideal candidates for cancer chemoprevention and therapy.

1. INTRODUCTION

Medicinal plants play a vital role in the discovery and development of drugs. The neem (*Azadirachta indica A. Juss*), an evergreen tree widely distributed in Asia, Africa, and America and especially ubiquitous in the Indian subcontinent, has attracted the focus of researchers globally due to its rich array of compounds with medicinal properties. Neem has been documented to exhibit a wide range of biological activities including antimicrobial, anti-inflammatory, immunostimulant, antioxidant, antigenotoxic, and anticancer activities. All parts of the neem tree have been used in traditional systems of medicine to treat various human diseases including cancer [1–5].

Of more than 300 bioactive compounds isolated and characterized from various parts of the neem tree, the tetranorterpenoids also known as limonoids are responsible for the medicinal properties of neem. The limonoids are highly oxygenated modified triterpenes with a 4,4,8-trimethyl-17 furanyl steroid skeleton that constitute one-third of the phytochemicals in neem. The structural features vital for biological activity include an intact apoeuphol skeleton, a 14,15,-β-epoxide, and a reactive site (19–28 lactol bridge or a cyclohexane) on the A ring [6–8].

The predominant neem limonoids include azadirachtin, nimbolide, nimbin, deacetylnimbin, mahmoodin, salannin, epoxy-azadiradione, deacetylgedunin, and gedunin (Fig. 7.1). Several neem limonoids have been documented to display potent anticancer effects (Table 7.1). Of these, azadirachtin, nimbolide, and gedunin have been more extensively investigated for anticancer activity relative to other limonoids. Azadirachtin, isolated from seed kernels is a C-secolimonoid containing a decalin and hydroxy furan acetal moieties, with 16 sterogenic centers, 7 of which are fully substituted. The limonoid gedunin is found abundantly in neem seeds. Nimbolide present in leaves and flowers contains a classical limonoid skeleton with α,β-unsaturated ketone structural element and a δ-lactonic ring. Recently, Manosroi *et al.* [24] identified six new limonoids, 7-benzoyl-17-epinimbocinol, 3-acetyl-7-tigloylnimbidinin, 1-isovaleroyl-1-detigloylsalanninolide, 2,3-dihydro-3α-methoxy-nimbolide, deacetyl-20,21-epoxy-20,22-dihydro-21-deoxyisonimbinolide, and deacetyl-20,21,22,

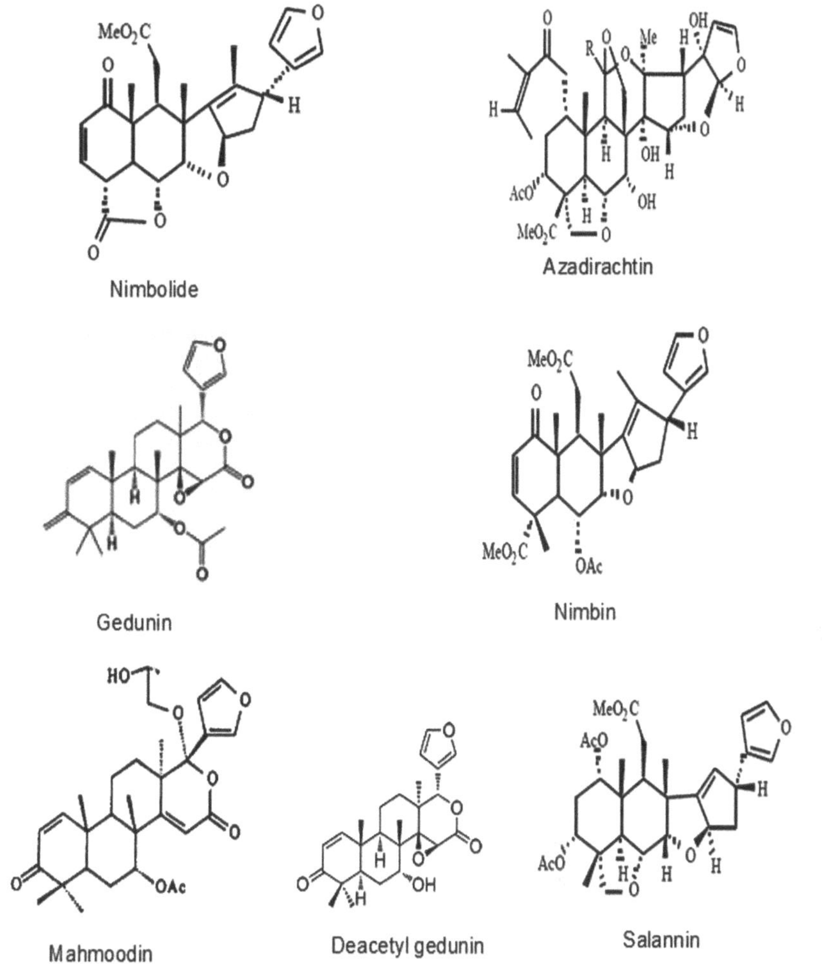

Figure 7.1 Structure of limonoids.

23-tetrahydro-20,22-dihydroxy-21,23-dimethoxynimbin, together with 28 known limonoids, 1–4, 6, 7, 9–14, 17–25, and 28–34 from the extracts of bark, leaves, roots, and seeds of *A. indica.* The neem limonoids have been shown to exert their anticancer effects by targeting key signaling pathways aberrant in cancer [7,25,26].

2. CYTOTOXICITY OF NEEM LIMONOIDS

Several neem limonoids have been demonstrated to exert cytotoxic effects against a diverse range of cancer cells in vitro. Kikuchi *et al.* [27] evaluated the cytotoxicity of 35 limonoids from *A. indica* seed extracts

Table 7.1 Key Neem Limonoids and Their Potential Targets

Limonoid	Target Genes	Function	Cancer Type	References
Azadirachtin	Cyclins, PCNA, P21	Antiproliferative	Cervical cancer	[9]
	PCNA, P21, cyclin D1		Oral cancer	Harish Kumar *et al.* (2010)
	p53, Fas, Bcl-2, Bax, Bid, Apaf-1, cytochrome *c*, survivin, caspases, and PARP	Apoptosis induction	Oral cancer	Harish Kumar *et al.* (2010)
	MMP-2, MMP-9	Anti-invasive	Oral cancer	[10]
	VEGF, VEGFR1, VEGFR2, HIF1α	Antiangiogenic	Oral cancer	[10]
Gedunin	Bionformatic analysis identified 532 genes involved	Antiproliferative	Ovarian cancer	[11]
	P23, caspase-7	Apoptosis induction	Breast cancer	[12]
Nimbolide	Cyclins, PCNA, P21	Antiproliferative	Breast cancer	[13]
	PCNA		Choriocarcinoma	[15]
	PCNA, P21, cyclin D1		Oral cancer	Harish Kumar *et al.* (2010)
	Cyclins, PCNA, P21		Cervical cancer	[9]
	Cyclins, CDKs, CDKIs, check point proteins		Colon cancer, leukemia	[16–18]
	Cyclin D		Colorectal cancer	[19]

Table 7.1 Key Neem Limonoids and Their Potential Targets—cont'd

Limonoid	Target Genes	Function	Cancer Type	References
	Bax, Bcl-2, cytochrome *c*, XIAP, caspases	Apoptosis induction	Prostate cancer	[20]
	Bcl-2, Bcl-xL, cIAP-1, survivin, and Mcl-1		Colorectal cancer	[19]
	Bcl-2, Bax, cytochrome *c*, Smac/DIABLO, caspases		Hepatocarcinoma	[21]
	Caspases		Colon cancer	[16]
	Bax, I-FLICE, cIAP-1, cIAP-2, Bcl-2, Bcl-xL, survivin, XIAP		Colon cancer	[22]
	Bcl-2, Bcl-xL, IAP-1, and IAP-2		Myeloma	[23]
	p53, Fas, Bcl-2, Bax, Bid, Apaf-1, cytochrome *c*, survivin, caspases, and PARP		Oral cancer	Harish Kumar *et al.* (2010)
	Bcl-2, Bax, Apaf 1, caspase-3, and PARP		Choriocarcinoma	[15]
	MMP-9, ICAM-1	Anti-invasive	Colorectal cancer	[19]
	MMP-2, MMP-9		Colon cancer	[16]
	MMP-9		Myeloma	[23]
	MMP-2, MMP-9		Oral cancer	[10]
	CXCR4	Antimetastatic	Colorectal cancer	[19]
	VEGF	Antiangiogenic	Colorectal cancer	[19]
	VEGF		Colon cancer	[16]
	VEGF		Myeloma	[23]
	VEGF, VEGFR1, VEGFR2, HIF1α		Oral cancer	[10]

against five human cancer cell lines. Three of these limonoids 7-deacetyl-7-benzoylepoxyazadiradione, 7-deacetyl-7-benzoylgeduin, and 28-deoxonimbolide exhibited selective cytotoxic effects against HL-60 leukemia cells. Azadirachtin was shown to exert cytotoxic effects against glioblastoma, human cervical cancer (HeLa), MCF-7 breast cancer, 143B TK human osteosarcoma, and ovarian cell lines [9,28,29]. Several studies have documented the antiproliferative effects of gedunin against various cancer cell lines including prostate, colon, ovarian, breast, nonsmall cell lung, melanoma and cervical cancer [11,12,27,30–33].

Nimbolide, the most potent neem limonoid was found to inhibit the growth of myriad cancer cell lines including 143B TK osteosarcoma, HL-60, U-937 and THP-1 leukemic, B16 melanoma, SMMC 7721, A-549, MCF-7 breast, MD-MBA-231, HT-29, SW-620, SW-480, HOP-62, A-549, prostate PC-3, choriocarcinoma (BeWo), leukemic (HeLa) and WiDr and HCT-116 colon adenocarcinoma and ovarian OVCAR-4,5,8 cell lines [9,13–17,34–36]. The cytotoxicity of neem limonoids has been attributed to their effects on the hallmarks of cancer.

3. NEEM LIMONOIDS AND HALLMARKS OF CANCER

The growth and functional capabilities of cancer cells to survive, proliferate, and disseminate are acquired through eight essential alterations referred to as *hallmarks of cancer*. These include self-sufficiency in growth signals, insensitivity to growth-inhibitory signals, evasion of programmed cell death, limitless replicative potential, sustained angiogenesis, tissue invasion and metastasis, reprogramming energy metabolism, and evading immune destruction. In addition, the expression of stress phenotypes such as oxidative stress, DNA damage/replication stress, proteotoxic stress, mitotic stress, and metabolic stress are regarded as additional hallmark capabilities of neoplastic cells. The acquisition of these hallmark capabilities is mediated through two enabling characteristics, genomic instability and inflammation [37]. Extensive studies by us and others have revealed that neem limonoids exert anticancer effects by modulating the hallmark capabilities of cancer (Fig. 7.2).

3.1. Inhibition of Cell Proliferation

Perturbation of cell cycle control and unrestricted cell proliferation are hallmark features of neoplasia. Progression of cells through the G1, S, G2, and

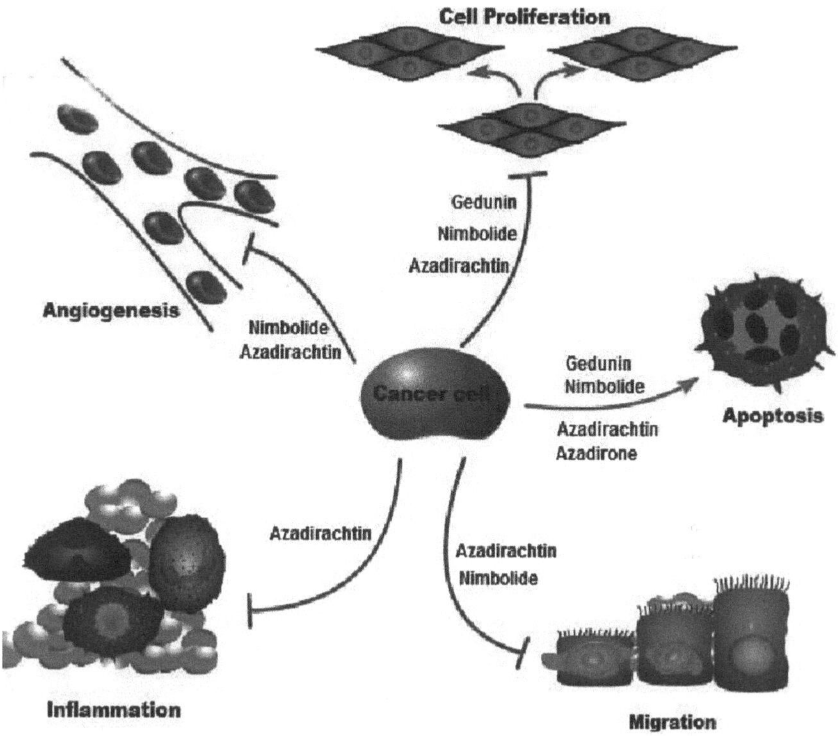

Figure 7.2 Neem limonoids inhibit hallmark capabilities of cancer. Azadirachtin, gedunin, and nimbolide selectively target cell proliferation, apoptosis, and inhibition of migration, inflammation, and angiogenesis. Azadirachtin and nimbolide are involved in inhibiting cell proliferation, migration, and angiogenesis and induce apoptosis in cancer cells. In addition, azadirachtin is involved in inhibiting inflammation. Gedunin is mainly involved in restraining cell proliferation and inducing apoptosis. (See the color plate.)

M phases of the cell cycle involves an intricate sequence of events regulated by cyclins, cyclin-dependent kinases (CDKs), and their inhibitors.

The cytotoxicity of azadirachtin has been attributed to increased micronuclei formation coupled to decreased mitotic index [28]. Furthermore, azadirachtin-induced cell cycle arrest at G2/M phase of cell cycle with accumulation of aberrant mitotic figures interfered with polymerization of tubules and formation of the mitotic spindle, thereby preventing replication. Accumulating evidence indicates that azadirachtin and nimbolide interfere with cell cycle kinetics in cancer cells by inducing cycle arrest at G1/S or G2/M phase through repression of cyclins, CDKs, PCNA, c-myc, and Rad17 accompanied by p53-dependent accumulation of p21Cip1/waf1 and Chk2 [9,16–18,24,38,39]. Gedunin was reported to arrest cell cycle in breast cancer, non small cell lung cancer, melanoma and ovarian cancer cell lines [11,30].

3.2. Apoptosis Induction

Apoptosis evasion, a major hallmark of human cancers promotes tumor development and progression as well as resistance to antineoplastic drugs. Tumor cells evade apoptosis by downregulation of death receptors, proapoptotic proteins, caspases, and upregulation of antiapoptotic proteins. Induction of apoptosis is one of the important therapeutic strategies to inhibit cancer cell proliferation. Extensive investigations have revealed that neem limonoids induce apoptosis by both the death receptor and the mitochondrial pathways [9,25,40,63].

Azadirone, a neem limonoid, sensitizes cancer cells to TNF-related apoptosis-inducing ligand (TRAIL) via ROS–ERK–CHOP-mediated upregulation of death receptor 4 and 5 signaling [19]. 7-Deacetyl-7-benzoylepoxyazadiradione, 7-deacetyl-7-benzoylgeduin, and 28-deoxonimbolide isolated from the seeds of *A. indica* were shown to induce both mitochondrial and death receptor-mediated apoptosis through activation of caspases-3, -8, and -9 in HL-60 cells. Gedunin induced apoptosis in HL-60 leukemia cells via both the mitochondrial and the death receptor-mediated pathways as evidenced by caspase activation and PARP cleavage. This was associated with inactivation of the heat-shock protein 90 chaperonin machinery [12,27]. Nimbolide induced various proapoptotic proteins while suppressing antiapoptotic proteins in estrogen-dependent (MCF-7) and estrogen-independent (MDA-MB-231) human breast cancer cell lines [13].

Studies from this laboratory provided evidence that azadirachtin and nimbolide target key molecules in both the intrinsic and extrinsic pathways to induce apoptosis. These limonoids transduced mitochondrial apoptosis by generation of reactive oxygen species, decline in the mitochondrial transmembrane potential with a shift of balance toward a proapoptotic phenotype through altered expression of Bcl-2 and IAP family of proteins, thereby stimulating cytochrome *c* efflux into the cytosol and triggering the caspase cascade [21]. Gupta *et al.* [22] demonstrated that nimbolide sensitizes HCT-116 and HT-29 human colon adenocarcinoma cells to TRAIL-induced apoptosis by upregulating the expression of DR4 and DR5 death receptors followed by recruitment of Fas-associated death domain and procaspase-8, and activation of executioner caspases eventually leading to cell death.

3.3. Effects on Tumor Invasion and Angiogenesis

Invasion and angiogenesis, critical events in tumor progression and metastasis depend on an intricate balance between proinvasive molecules such as matrix metalloproteinases (MMPs) involved in extracellular matrix (ECM) processing, proangiogenic factors predominantly vascular endothelial growth factor (VEGF), and their inhibitors. Studies by us and others have shown that nimbolide prevents tumor invasion and angiogenesis by downregulating the expression of MMPs, placental growth factor (PlGF), VEGF, VEGF receptors, and HIF-1α and upregulating TIMP-2 and RECK expression [10].

3.4. Anti-Inflammatory Effects

Many tumors arise from sites of chronic irritation and inflammation. Inflammation is mediated by inflammatory cells, which produce bioactive molecules such as growth factors, cytokines, and chemokines which involve in development of cancer. In addition, inflammation contributes to other hallmarks by providing growth factors, proangiogenic factors, and ECM-modifying enzymes that facilitate angiogenesis, invasion, and metastasis [37,41].

Neem limonoids exert anti-inflammatory effects by inhibiting the pro-inflammatory cytokine tumor necrosis factor (TNF)-α, as well as NF-κB that play a pivotal role in inflammation. Azadirachtin has been shown to block TNF-mediated inflammatory response via modulation of cell surface TNF receptors, inhibit NF-κB activation and downregulate the expression of the pro-inflammatory mediator Cox-2 [42]. The anti-inflammatory effects of epoxyazadiradione are mediated by preventing the release of the pro-inflammatory cytokines TNF-α and interleukin (IL) -1α, -1β, and -6, inhibition of NF-κB nuclear translocation and stimulation of inducible nitric oxide synthase and nitric oxide production [43]. Other neem limonoids exhibiting potent anti-inflammatory activity include 1,3-diacetylvilasinin, 28-deoxonimbolide, salannin, 2′,3′-dihydrosalanin, 3-deacetylsalanin [44].

3.5. Immunomodulatory Effects

Enhanced production and secretion of pro-inflammatory cytokines can cause immunosuppression that can facilitate tumor development and progression. Neem leaf extracts rich in limonoids have been reported to enhance antitumor immunity, activate cytotoxic T lymphocytes and natural killer cells and increase interferon-γ and IL-12 secretion [45–48].

Gedunin was reported to reduce CD69 and CD25 T lymphocytes and impair allergic responses [49].

3.6. Antioxidant Activity

Oxidative stress defined as a state of imbalance between the concentrations of ROS and the antioxidant defense mechanisms is one of the most important stress phenotypes implicated in malignant transformation. The neem limonoids azadirachtin and nimbolide exhibited significant protective effects against various free radicals and oxidative DNA damage by upregulation of antioxidants [10,50]. Gedunin was shown to reduce oxidative stress by activating Nrf2/Keap-1 signaling [51].

4. ONCOGENIC SIGNALING

Aberrant alteration of signaling pathways leads to acquisition of the hallmark capabilities of cancer. Among the plethora of molecules involved in cellular signaling, the upstream signaling kinases and downstream transcription factors play a central role in signal transduction pathways [52,53]. While the kinases catalyze phosphorylation of proteins, the transcription factors recognize specific DNA sequences and regulate transcription initiation either positively or negatively. The human genome contains 518 putative proteins and lipid kinases [54–56]. Both kinases and transcription factors regulate various fundamental biological processes such as proliferation, apoptosis, differentiation, cell adhesion, and migration.

Aberrant regulation of kinases including PI3K/Akt/mToR, RAS/RAF/MEK/ERK, and transcription factors such as β-catenin, NF-κB, and signal transducer and activator of transcription (STAT) contributes to neoplastic transformation [52,53]. The neem limonoids have been demonstrated to target oncogenic kinases and transcription factors to arrest tumor development and progression.

4.1. NF-κB Signaling

The transcription factor NF-κB is one of the most important targets of neem extracts as well as neem limonoids. In resting cells, NF-κB is found as an inactive heterodimer of p50 and p65 subunits sequestered to the inhibitory protein IκB. Phosphorylation of IκB at serine-32 and -36 residues by IKKβ leads to proteasomal degradation of IκB, followed by nuclear translocation of NF-κB and transactivation of genes involved in acquisition of the hallmark capabilities of cancer such as cell proliferation, cell survival, apoptosis evasion, invasion, metastasis, and angiogenesis [57].

Studies from this laboratory in human cancer cells *in vitro* and in the hamster model of oral oncogenesis *in vivo* provided evidence that azadirachtin and nimbolide abrogate NF-κB signaling by inhibiting the kinase activity of IKKβ, thereby preventing IκBα phosphorylation and degradation, NF-κB nuclear translocation, and transactivation of antiapoptotic genes [9,21,64]. In HepG2 cells, nimbolide blocked both inducible as well as constitutive activation of NF-κB as revealed by attenuation of TNF-induced activation of NF-κB [21]. In colon adenocarcinoma cells, nimbolide was found to restrain nuclear translocation and DNA binding of NF-κB by suppressing the phosphorylation of IκB-α via interaction with IKK [16], Gupta *et al.* [23] reported that nimbolide inhibits the kinase activity of IKK-β by modification of a critical cys (179) residue with consequent inactivation of NF-κB.

4.2. Wnt/β-Catenin Signaling

NF-κB activation is recognized to synergistically induce activation of canonical Wnt/β-catenin signaling pathways [58]. Activation of the Wnt/β-catenin pathway involves disruption of a multiprotein complex containing glycogen synthase kinase 3β (GSK-3β). This leads to cytosolic accumulation and nuclear translocation of β-catenin and transactivation of target genes involved in cell proliferation, apoptosis, epithelial–mesenchymal transition, invasion, and angiogenesis [59]. In HepG2 cells, nimbolide restrained canonical Wnt/β-catenin signaling by downregulating GSK-3β, and preventing the nuclear translocation of β-catenin [21].

4.3. PI3K/Akt Signaling

The phosphoinositide 3-kinases (PI3Ks) are lipid kinases that participate in the regulation of a variety of cellular processes including cell adhesion, cell cycle progression, cell migration, cell survival, differentiation, metabolism, proliferation, and transcription. Activation of Akt, a serine kinase and the main downstream target of PI3K leads to cell proliferation, survival, and angiogenesis. Nimbolide was shown to inhibit IGF-induced PI3K/Akt signaling in MCF-7 and MD-MBA-231 breast cancer cell lines [13]. Raja Singh *et al.* [20] demonstrated that nimbolide abrogates the PI3K/Akt pathway by downregulating PI3K, Akt, IGF-1, and IGFR in prostate cancer cell lines. In glioblastoma, nimbolide was shown to inhibit Akt phosphorylation both *in vitro* and *in vivo* [38]. Gedunin was also found to inhibit the PI3K/Akt and MAPK pathways [11].

4.4. MAPK Signaling

PI3K regulates its biological functions by activating transcription factors and the MAPK signaling pathways. The MAPK/ERK signaling cascade comprises extracellular signal regulated kinase (ERK), c-Jun N-terminal kinase (JNK), and p38 kinases which play a key role in transmitting extracellular signals into the cell. Activation of Ras, raf/MEK/ERK cascade induces target genes that are involved in various cellular processes [60]. Nimbolide abrogates the Ras/Raf/ERK and p38 pathway to inhibit cell proliferation and induce apoptosis in MCF-7 and MD-MBA-231 breast cancer cell lines [13]. Babykutty *et al.* [16] provided evidence to demonstrate that nimbolide induces caspase-mediated apoptosis by inhibiting ERK1/2 and activating p38 and JNK1/2 in colon cancer cells. Recently, nimbolide was reported to significantly downregulate ERK1/2 levels in glioblastoma multiforme cells *in vitro* and tumor xenografts [38]. Gedunin was found to inhibit RANKL induced osteoclastogenesis by downregulating p38, ERK, and NF-κBp65 [61].

4.5. JAK/STAT Signaling

STAT proteins are a family of cytoplasmic transcription factors that mediate normal cellular responses of cytokines, growth factors, and other polypeptide ligands [62]. Phosphorylation of STAT leads to dimerization, consequent nuclear translocation, and transactivation of target genes. STAT3 regulates the transcription of multiple key genes involved in cell proliferation, apoptosis, angiogenesis, invasion, and metastasis. Recently, Karkare *et al.* [38] reported that nimbolide inhibits STAT-3 pathway in glioblastoma cells.

5. CONCLUSIONS AND FUTURE PERSPECTIVES

With increasing knowledge about the aberrations in multiple signaling pathways in neoplastic transformation, the current focus in chemoprevention research is to identify agents that are capable of targeting molecules and signaling pathways involved in tumorigenesis. Neem limonoids such as azadirachtin, gedunin, and nimbolide target multiple cellular and molecular pathways that are aberrant in cancer including cell proliferation, apoptosis evasion, inflammation, invasion, and angiogenesis (Fig. 7.3). However,

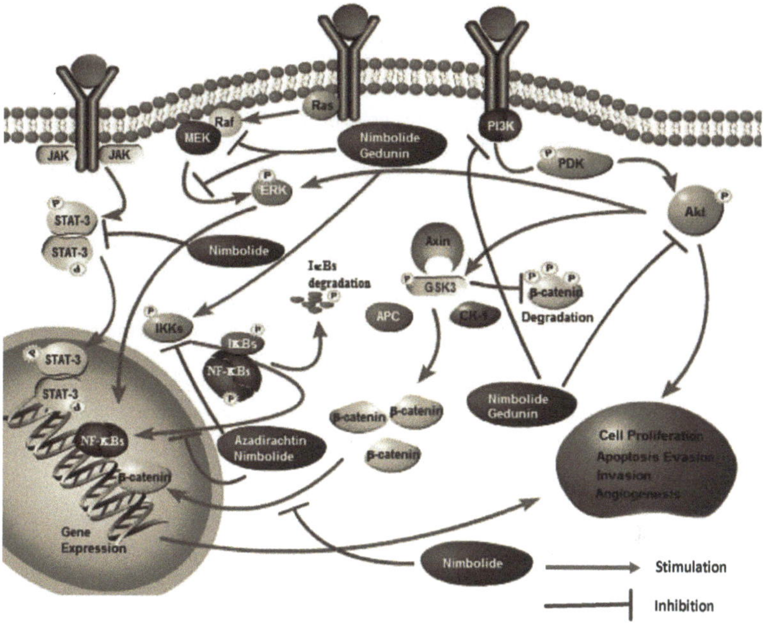

Figure 7.3 Schematic representation of the potential targets of azadirachtin, gedunin, and nimbolide. Limonoids target a plethora of molecules involved in signaling pathways. Neem limonoids mainly abrogate upstream signaling kinases and transcription factors. Azadirachtin and nimbolide inhibit IKKs, preventing degradation of IκB-α and nuclear translocation of NF-κB, thereby inhibiting the activation of NF-κB signaling and its downstream target genes. Nimbolide also inhibits Wnt/β-catenin signaling by blocking the accumulation and nuclear translocation of β-catenin. In addition, nimbolide also inhibits STAT-3 signaling. Gedunin and nimbolide are actively involved in inhibiting PI3K/Akt pathway by preventing activation of PI3K and Akt, thereby inhibiting cell proliferation, invasion, and angiogenesis. Furthermore, MAPK/ERK pathway is effectively inhibited by gedunin and nimbolide. (See the color plate.)

the efficacy of these limonoids has been tested only at the preclinical level and the effects of these phytochemicals on humans are largely unexplored. Extensive investigations on absorption, metabolism, tissue availability, toxicity, excretion, mechanism of action, crosstalk with other signaling pathways, and well-designed clinical trials are required for translating the beneficial effects of neem limonoids from the bench to the bedside.

REFERENCES

[1] A.E. Abdel Moneim, M.S. Othman, A.M. Aref, *Azadirachta indica* attenuates cisplatin-induced nephrotoxicity and oxidative stress, Biomed. Res. Int. 2014 (2014) 647131.

[2] K. Biswas, I. Chattopadhyay, R.K. Banerjee, Biological activities and medicinal properties of neem (*Azadirachta indica*), Curr. Sci. 82 (2002) 1336–1345.

[3] H. Kato-Noguchi, M.A. Salam, O. Ohno, K. Suenaga, Nimbolide B and nimbic acid B, phytotoxic substances in neem leaves with allelopathic activity, Molecules 19 (2014) 6929–6940.
[4] C. Sharma, A.J. Vas, P. Goala, T.M. Gheewala, T.A. Rizvi, A. Hussain, Ethanolic neem (*Azadirachta indica*) leaf extract prevents growth of MCF-7 and HeLa cells and potentiates the therapeutic index of cisplatin, J. Oncol. 2014 (2014) 321754.
[5] R. Subapriya, S. Nagini, Medicinal properties of neem leaves: A review, Curr. Med. Chem. Anticancer Agents 5 (2005) 149–156.
[6] D.E. Champagne, O. Koul, M.B. Isman, G.G.E. Scudder, G.H. Neil Towers, Biological activity of limonoids from the rutales, Phytochemistry 31 (1992) 377–394.
[7] F. Hao, S. Kumar, N. Yadav, D. Chandra, Neem components as potential agents for cancer prevention and treatment, Biochim. Biophys. Acta 1846 (2014) 247–257.
[8] E.G. Miller, J.L. Porter, W.H. Binnie, I.Y. Guo, S. Hasegawa, Further studies on the anticancer neem seed insecticide preparations, Pest Sci. 48 (2004) 135–140.
[9] R.V. Priyadarsini, R.S. Murugan, P. Sripriya, D. Karunagaran, S. Nagini, The neem limonoids azadirachtin and nimbolide induce cell cycle arrest and mitochondria-mediated apoptosis in human cervical cancer (HeLa) cells, Free Radic. Res. 44 (2010) 624–634.
[10] R.V. Priyadarsini, P. Manikandan, G.H. Kumar, S. Nagini, The neem limonoids azadirachtin and nimbolide inhibit hamster cheek pouch carcinogenesis by modulating xenobiotic-metabolizing enzymes, DNA damage, antioxidants, invasion and angiogenesis, Free Radic. Res. 43 (2009) 492–504.
[11] S.G. Kamath, N. Chen, Y. Xiong, R. Wenham, S. Apte, M. Humphrey, J. Cragun, J.M. Lancaster, Gedunin, a novel natural substance, inhibits ovarian cancer cell proliferation, Int. J. Gynecol. Cancer 19 (9) (2009) 1564–1569.
[12] C.A. Patwardhan, A. Fauq, L.B. Peterson, C. Miller, B.S. Blagg, A. Chadli, Gedunin inactivates the co-chaperone p23 protein causing cancer cell death by apoptosis, J. Biol. Chem. 288 (2013) 7313–7325.
[13] P. Elumalai, R. Arunkumar, C.S. Benson, G. Sharmila, J. Arunakaran, Nimbolide inhibits IGF-I-mediated PI3K/Akt and MAPK signalling in human breast cancer cell lines (MCF-7 and MDA-MB-231), Cell Biochem. Funct. 32 (2014) 476–484.
[14] G. Harish Kumar, K.V.P. Chandra Mohan, A.J. Rao, S. Nagini, Nimbolide a limonoid from human cervical cancer (HeLa) cells, Free Radic. Res. 44 (2009) 624–634.
[15] G. Harish Kumar, K.V.P. Chandra Mohan, A.J. Rao, S. Nagini, Nimbolide a limonoid from *Azadirachta indica* inhibits proliferation and induces apoptosis of human choriocarcinoma (BeWo) cells, Invest. New Drugs 27 (2009) 246–252.
[16] S. Babykutty, S. PP, J. NR, M.A. Kumar, M.S. Nair, P. Srinivas, S. Gopala, Nimbolide retards tumor cell migration, invasion, and angiogenesis by downregulating MMP-2/9 expression via inhibiting ERK1/2 and reducing DNA-binding activity of NF-κB in colon cancer cells, Mol. Carcinog. 51 (2012) 475–490.
[17] M.K. Roy, M. Kobori, M. Takenaka, K. Nakahara, H. Shinmoto, S. Isobe, T. Tsushida, Antiproliferative effect on human cancer cell lines after treatment with nimbolide extracted from an edible part of the neem tree (*Azadirachta indica*), Phytother. Res. 21 (2007) 245–250.
[18] M.K. Roy, M. Kobori, M. Takenaka, K. Nakahara, H. Shinmoto, T. Tsushida, Inhibition of colon cancer (HT-29) cell proliferation by a triterpenoid isolated from *Azadirachta indica* is accompanied by cell cycle arrest and up-regulation of p21, Planta Med. 72 (2006) 917–923.
[19] S.C. Gupta, S.K. Francis, M.S. Nair, Y.Y. Mo, B.B. Aggarwal, Azadirone, a limonoid tetranortriterpene, induces death receptors and sensitizes human cancer cells to tumor necrosis factor-related apoptosis-inducing ligand (TRAIL) through a p53

protein-independent mechanism: Evidence for the role of the ROS-ERK-CHOP-death receptor pathway, J. Biol. Chem. 288 (2013) 32343–32356.
[20] P. Raja Singh, R. Arunkumar, V. Sivakamasundari, G. Sharmila, P. Elumalai, E. Suganthapriya, A. Brindha Mercy, K. Senthilkumar, J. Arunakaran, Anti-proliferative and apoptosis inducing effect of nimbolide by altering molecules involved in apoptosis and IGF signalling via PI3K/Akt in prostate cancer (PC-3) cell line, Cell Biochem. Funct. 32 (2014) 217–228.
[21] K. Kavitha, R. Vidya Priyadarsini, P. Anitha, K. Ramalingam, R. Sakthivel, G. Purushothaman, A.K. Singh, D. Karunagaran, S. Nagini, Nimbolide, a neem limonoid abrogates canonical NF-κB and Wnt signaling to induce caspase dependent apoptosis in human hepatocarcinoma HepG2 cells, Eur. J. Pharmacol. 681 (2012) 6–14.
[22] S.C. Gupta, S. Reuter, K. Phromnoi, B. Park, P.S. Hema, M. Nair, B.B. Aggarwal, Nimbolide sensitizes human colon cancer cells to TRAIL through reactive oxygen species- and ERK-dependent up-regulation of death receptors, p53, and Bax, J. Biol. Chem. 286 (2011) 1134–1146.
[23] S.C. Gupta, S. Prasad, S. Reuter, R. Kannappan, V.R. Yadav, J. Ravindran, P.S. Hema, M.M. Chaturvedi, M. Nair, B.B. Aggarwal, Modification of cysteine 179 of IkappaB-alpha kinase by nimbolide leads to down-regulation of NF-kappaB-regulated cell survival and proliferative proteins and sensitization of tumor cells to chemotherapeutic agents, J. Biol. Chem. 285 (2010) 35406–35417.
[24] A. Manosroi, W. Kitdamrongtham, K. Ishii, T. Shinozaki, Y. Tachi, M. Takagi, K. Ebina, J. Zhang, J. Manosroi, R. Akihisa, T. Akihisa, Limonoids from *Azadirachta indica* var. siamensis extracts and their cytotoxic and melanogenesis-inhibitory activities, Chem. Biodivers. 11 (2014) 505–531.
[25] L.N. Bodduluru, E.R. Kasala, N. Thota, C.C. Barua, R. Sistla, Chemopreventive and therapeutic effects of nimbolide in cancer: The underlying mechanisms, Toxicol. In Vitro 28 (2014) 1026–1035.
[26] M.J. Gualtieri, N. Malafronte, A. Vassallo, A. Braca, R. Cotugno, M. Vasaturo, N. De Tommasi, F. Dal Piaz, Bioactive limonoids from the leaves of *Azaridachta indica* (Neem), J. Nat. Prod. 77 (2014) 596–602.
[27] T. Kikuchi, K. Ishii, T. Noto, A. Takahashi, K. Tabata, T. Suzuki, T. Akihisa, Cytotoxic and apoptosis-inducing activities of limonoids from the seeds of *Azadirachta indica* (neem), J. Nat. Prod. 74 (2011) 866–870.
[28] J. Akudugu, G. Gäde, L. Böhm, Cytotoxicity of azadirachtin A in human glioblastoma cell lines, Life Sci. 68 (2001) 1153–1160.
[29] E. Cohen, G.B. Quistad, J.E. Casida, Cytotoxicity of nimbolide, epoxyazadiradione and other limonoids from neem insecticide, Life Sci. 58 (1996) 1075–1081.
[30] C.M. Cazal, K. Choosang, V.G. Severino, M.S. Soares, A.L. Sarria, J.B. Fernandes, M.F. Silva, P.C. Vieira, P. Pakkong, G.M. Almeida, M.H. Vasconcelos, M.S. Nascimento, M.M. Pinto, Evaluation of effect of triterpenes and limonoids on cell growth, cell cycle and apoptosis in human tumor cell line, Anticancer Agents Med Chem. 10 (2010) 769–776.
[31] H. Hieronymus, J. Lamb, K.N. Ross, X.P. Peng, C. Clement, A. Rodina, M. Nieto, J. Du, K. Stegmaier, S.M. Raj, K.N. Maloney, J. Clardy, W.C. Hahn, G. Chiosis, T.R. Golub, Gene expression signature-based chemical genomic prediction identifies a novel class of HSP90 pathway modulators, Cancer Cell 10 (2006) 321–330.
[32] J. Lamb, E.D. Crawford, D. Peck, J.W. Modell, I.C. Blat, M.J. Wrobel, J. Lerner, J.P. Brunet, A. Subramanian, K.N. Ross, M. Reich, H. Hieronymus, G. Wei, S.A. Armstrong, S.J. Haggarty, P.A. Clemons, R. Wei, S.A. Carr, E.S. Lander, T.R. Golub, The Connectivity Map: Using gene-expression signatures to connect small molecules, genes, and disease, Science 313 (2006) 1929–1935.

[33] S.J. Uddin, L. Nahar, J.A. Shilpi, M. Shoeb, T. Borkowski, S. Gibbons, M. Middleton, M. Byres, S.D. Sarker, Gedunin, a limonoid from *Xylocarpus granatum*, inhibits the growth of CaCo-2 colon cancer cell line in vitro, Phytother. Res. 21 (2007) 757–761.
[34] J. Chen, J. Chen, Y. Sun, Y. Yan, L. Kong, Y. Li, M. Qiu, Cytotoxic triterpenoids from *Azadirachta indica*, Planta Med. 77 (2011) 1844–1847.
[35] E. Cohen, G.B. Quisted, P.R. Jefferies, Nimbolide is the principle cytotoxic component of neem seed insecticide preparations, Pest Sci. 48 (1996) 135–140.
[36] B.S. Sastry, K. Suresh Babu, T. Hari Babu, S. Chandrasekhar, P.V. Srinivas, A.K. Saxena, M.K. Madhusudana Roy, M. Kobori, M. Takenaka, K. Nakahara, H. Shinmoto, S. Isobe, T. Tsushida, Antiproliferative effect on human cancer cell lines after treatment with nimbolide extracted from an edible part of the neem tree (*Azadirachta indica*), Phytother. Res. 21 (2007) 245–250.
[37] D. Hanahan, R.A. Weinberg, Hallmarks of cancer: The next generation, Cell 144 (2011) 646–674.
[38] S. Karkare, R.R. Chhipa, J. Anderson, X. Liu, H. Henry, A. Gasilina, N. Nassar, J. Roychoudhury, J.P. Clark, A. Kumar, G.M. Pauletti, P.K. Ghosh, B. Dasgupta, Direct inhibition of retinoblastoma phosphorylation by nimbolide causes cell-cycle arrest and suppresses glioblastoma growth, Cancer Res. 20 (2014) 199–212.
[39] M. Takagi, Y. Tachi, J. Zhang, T. Shinozaki, K. Ishii, T. Kikuchi, M. Ukiya, N. Banno, H. Tokuda, T. Akihisa, Cytotoxic and melanogenesis-inhibitory activities of limonoids from the leaves of *Azadirachta indica* (neem), Chem. Biodivers. 11 (2014) 451–468.
[40] P. Srivastava, N. Yadav, R. Lella, A. Schneider, A. Jones, T. Marlowe, G. Lovett, K. O'Loughlin, H. Minderman, R. Gogada, D. Chandra, Neem oil limonoids induces p53-independent apoptosis and autophagy, Carcinogenesis 33 (2012) 2199–2207.
[41] L.M. Coussens, Z. Werb, Inflammation and cancer, Nature 420 (2002) 860–867.
[42] M. Thoh, P. Kumar, H.A. Nagarajaram, S.K. Manna, Azadirachtin interacts with the tumor necrosis factor (TNF) binding domain of its receptors and inhibits TNF-induced biological responses, J. Biol. Chem. 285 (2010) 5888–5895.
[43] A. Alam, S. Haldar, H.V. Thulasiram, R. Kumar, M. Goyal, M.S. Iqbal, C. Pal, S. Dey, S. Bindu, S. Sarkar, U. Pal, N.C. Maiti, U. Bandyopadhyay, Novel anti-inflammatory activity of epoxyazadiradione against macrophage migration inhibitory factor: Inhibition of tautomerase and proinflammatory activities of macrophage migration inhibitory factor, J. Biol. Chem. 287 (2012) 24844–24861.
[44] T. Akihisa, A. Takahashi, T. Kikuchi, M. Takagi, K. Watanabe, M. Fukatsu, Y. Fujita, N. Banno, H. Tokuda, K. Yasukawa, The melanogenesis-inhibitory, anti-inflammatory, and chemopreventive effects of limonoids in n-hexane extract of *Azadirachta indica A. Juss*. (neem) seeds, J. Oleo Sci. 60 (2011) 53–59.
[45] K. Chakraborty, A. Bose, T. Chakraborty, K. Sarkar, S. Goswami, S. Pal, R. Baral, Restoration of dysregulated CC chemokine signaling for monocyte/macrophage chemotaxis in head and neck squamous cell carcinoma patients by neem leaf glycoprotein maximizes tumor cell cytotoxicity, Cell. Mol. Immunol. 7 (2010) 396–408.
[46] T. Chakraborty, A. Bose, S. Barik, K.K. Goswami, S. Banerjee, S. Goswami, D. Ghosh, S. Roy, K. Chakraborty, K. Sarkar, R. Baral, Neem leaf glycoprotein inhibits CD4+CD25+Foxp3+ Tregs to restrict murine tumor growth, Immunotherapy 3 (2011) 949–969.
[47] F.H. Sarkar, Y. Li, NF-kappaB: A potential target for cancer chemoprevention and therapy, Front. Biosci. 13 (2008) 2950–2959.
[48] K. Sarkar, S. Goswami, S. Roy, A. Mallick, K. Chakraborty, A. Bose, R. Baral, Neem leaf glycoprotein enhances carcinoembryonic antigen presentation of dendritic cells to T and B cells for induction of anti-tumor immunity by allowing generation of immune effector/memory response, Int. Immunopharmacol. 10 (2010) 865–874.

[49] F.K. Ferraris, K.H. Moret, A.B. Figueiredo, C. Penido, Henriques Md, Gedunin, a natural tetranortriterpenoid, modulates T lymphocyte responses and ameliorates allergic inflammation, Int. Immunopharmacol. 14 (2012) 82–93.
[50] P. Manikandan, R. Anandan, S. Nagini, Evaluation of *Azadirachta indica* leaf fractions for *in vitro* antioxidant potential and protective effects against H_2O_2-induced oxidative damage to pBR322 DNA and red blood cells, J. Agric. Food Chem. 57 (2009) 6990–6996.
[51] N.A. Smirnova, R.E. Haskew-Layton, M. Basso, D.M. Hushpulian, J.B. Payappilly, R.E. Speer, Y.H. Ahn, I. Rakhman, P.A. Cole, J.T. Pinto, R.R. Ratan, I.G. Gazaryan, Development of Neh2-luciferase reporter and its application for high throughput screening and real-time monitoring of Nrf2 activators, Chem. Biol. 18 (2011) 752–765.
[52] G.S. Martin, Cell signaling and cancer, Cancer Cell 4 (2003) 167–174.
[53] T. Pawson, N. Warner, Oncogenic re-wiring of cellular signaling pathways, Oncogene 26 (2007) 1268–1275.
[54] K. Deshmukh, K. Anamika, N. Srinivasan, Evolution of domain combinations in protein kinases and its implications for functional diversity, Prog. Biophys. Mol. Biol. 102 (2010) 1–15.
[55] G. Manning, The protein kinase complement of the human genome, Science 298 (2002) 1912–1934.
[56] J. Rauch, N. Volinsky, D. Romano, W. Kolch, The secret life of kinases: Functions beyond catalysis, Cell Commun. Signal. 28 (2011) 1–28.
[57] M.M. Chaturvedi, B. Sung, V.R. Yadav, R. Kannappan, B.B. Aggarwal, NF-κB addiction and its role in cancer: 'one size does not fit all', Oncogene 30 (2011) 1615–1630.
[58] Q. Du, D.A. Geller, Cross-regulation between Wnt and NF-κB signaling pathways, Forum Immunopathol. Dis. Therap. 1 (2010) 155–181.
[59] J.K. Sethi, A. Vidal-Puig, Wnt signalling and the control of cellular metabolism, Biochem. J. 427 (2010) 1–17.
[60] F.F. Xu, H. Zhu, X.M. Li, F. Yang, J.D. Chen, B. Tang, H.G. Sun, Y.N. Chu, R.X. Zheng, Y.L. Liu, L.S. Wang, Y. Zhang, Intercellular adhesion molecule-1 inhibits osteogenic differentiation of mesenchymal stem cells and impairs bio-scaffold-mediated bone regeneration in vivo, Tissue Eng. Part A 20 (2014) 2768–2782.
[61] C. Wisutsitthiwong, C. Buranaruk, K. Pudhom, T. Palaga, The plant limonoid 7-oxo-deacetoxygedunin inhibits RANKL-induced osteoclastogenesis by suppressing activation of the NF-κB and MAPK pathways, Biochem. Biophys. Res. Commun. 415 (2011) 361–366.
[62] S. Yan, Z. Li, C.J. Thiele, Inhibition of STAT3 with orally active JAK inhibitor, AZD1480, decreases tumor growth in neuroblastoma and pediatric sarcomas in vitro and in vivo, Oncotarget 4 (2013) 433–445.
[63] P. Elumalai, D.N. Gunadharini, K. Senthilkumar, S. Banudevi, R. Arunkumar, C.S. Benson, G. Sharmila, J. Arunakaran, Induction of apoptosis in human breast cancer cells by nimbolide through extrinsic and intrinsic pathway, Toxicol. Lett. 215 (2012) 131–142.
[64] G. Harish Kumar, R. Vidya Priyadarsini, G. Vinothini, P. Vidjaya Letchoumy, S. Nagini, The neem limonoids azadirachtin and nimbolide inhibit cell proliferation and induce apoptosis in an animal model of oral oncogenesis, Invest. New Drugs 28 (2010) 392–401.

CHAPTER EIGHT

Curcumin: A Potent Modulator of Multiple Enzymes in Multiple Cancers

Adeeb Shehzad*, Raheem Shahzad†, Young Sup Lee*,[1]

*School of Life Sciences, BK21 Plus KNU Creative BioResearch Group, Kyungpook National University, Daegu, Republic of Korea

†School of Applied Biosciences, College of Agriculture and Life Sciences, Kyungpook National University, Daegu, Republic of Korea

[1]Corresponding author: e-mail address: yselee@knu.ac.kr

Contents

The Enzymes, Volume 36
ISSN 1874-6047
http://dx.doi.org/10.1016/B978-0-12-802215-3.00008-2

Abstract

Curcumin (diferuloylmethane) is the biphenolic active compound of turmeric. Curcumin has been used for hundreds of years to treat various ailments. Curcumin has been reported to exert numerous pharmacological effects by modulating multiple molecular targets including those involved in the pathogenesis of cancer. Cancer has been characterized as the dysregulation of cell signaling pathways through gradual alteration of regulatory proteins and through gene mutation. Curcumin is a highly pleiotropic molecule that modulates several intracellular signaling pathways in cancer. The pleiotropic activities of curcumin have been attributed to its novel molecular structure. Based on its β-diketone moiety, curcumin exists in keto–enol tautomers, and this tautomerism favors interaction and binding with a wide range of enzymes. Several studies have shown modulation of numerous signaling enzymes by curcumin including, LOX, COX-2, XO, proteasomes, Ca^{2+}-ATPase of sarcoplasmic reticulum, MMPs, HAT, HDAC, DNMT1, DNA polymerase λ, ribonucleases, GloI, protein kinases (PKA, PKB, PKC, v-Src, GSK-3β, ErbB2), protein reductases (TrxR1, AR), GSH, ICDHs, peroxidases (Prx1, Prx2, Prx6) by treatment with curcumin. Various biophysical analyses have been reported, which shows the underlying molecular interaction of curcumin with multiple targets in terms of binding affinities. The current chapter describes how curcumin binds and modulates multiple enzymes involved cancer. Published clinical trial studies with curcumin in cancer management will also be discussed.

1. INTRODUCTION

Curcumin is a yellow compound isolated from the dried ground rhizomes of the perennial herb *Curcuma longa* of the family Zingiberaceae. Curcumin has been used in traditional medicine for the treatment of different ailments including biliary disorders and inflammatory diseases [1]. Curcumin has been used as a dietary supplement in Asian countries and has not been found to have any major toxicity [1]. The pleiotropic role of this dietary agent includes the inhibition of several cell signaling pathways [2]. There is compelling evidence that curcumin-modulated enzymes, growth factors and their receptors, cytokines, and various protein kinases (PKs) that control cell proliferation and cell cycle progression [2]. Therefore, curcumin may have potential therapeutic value for digestive system, urological reproductive system, hematological cancers, and other tumors [3]. The optimal daily intake of curcumin is 0–1 mg/kg according to the Food and Agriculture Organization and the World Health Organization [4]. It has been demonstrated in clinical studies that curcumin is nontoxic and well tolerated, at doses up to 12 g/day [5]. Currently, several clinical trials are in progress to evaluate curcumin anticancer effects in various cancers (https://clinicaltrials.gov).

The molecular basis of cancer has been characterized by dysregulation of an array of signaling molecules. Cancer is a major public health problem in many parts of the world. American Cancer Society reported that a total of 1,665,540 new cases of cancer (855,220 male and 810,320 female) as well as 585,720 deaths due to cancer (310,010 male and 275,710 female) are expected in the United States in 2014 [6]. In the last decade, the rapid increase in the cost of health care as well as limited effectiveness of single target cancer therapies has facilitated a growing interest in phytochemicals for use in the prevention and treatment of cancer [7]. Curcumin has been reported to directly or indirectly modulate the activity of various enzymes by covalent and noncovalent interactions and hydrogen bonding [8]. Curcumin binds and modulates the activity of various enzymes such as lipoxygenases (LOXs), cyclooxygenase-2 (COX-2), xanthine oxidase (XO), proteasomes, Ca^{2+}-ATPase of sarcoplasmic reticulum, matrix metalloproteinases (MMPs), histone acetyltransferase (HAT), histone deacetylase (HDAC), DNMT1, DNA polymerase λ, ribonucleases (RNases), glyoxalases I (GloI), protein kinases (PKA, PKB, PKC, viral sarcoma (v-Src), glycogen synthase kinase-3 beta (GSK-3β), v-erb-b2 avian erythroblastic leukemia viral oncogene homolog 2 (ErbB2), protein reductases (thioredoxin reductase 1 (TrxR1), aldose reductase (AR)), GSH, isocitrate dehydrogenases (ICDHs), peroxidases (Prx1, Prx2, Prx6).

Curcumin has been found to block the activation of cytochrome P450 isozymes. Curcumin also enhanced the expression of phase I and II carcinogen-detoxifying enzymes, which may be responsible for its cancer chemopreventive effects [9]. Several studies have shown that curcumin inhibited the proliferation of various cancer cell lines, including those resistant to cancer chemotherapy [10]. Curcumin has the ability to suppress the abnormal expression of cyclin D1 in various cancers. Cyclin D1 is a subunit of cyclin-dependent kinase 4 (Cdk4) and Cdk6, which drive the movement of cells during cell cycle transitions [11]. Curcumin also blocked inflammation-induced cancer by downregulating the expression of COX-2, MMP9, cyclin D1, and adhesion molecules [11]. Several *in vitro*, *in vivo*, and clinical studies have shown that curcumin interacts with multiple enzymes and has the potential to inhibit the initiation and development of a wide variety of cancers. The binding efficiency and interaction of curcumin with multiple enzymes was determined using biophysical analyses such as molecular docking [12]. Additional studies such as X-ray crystallography and nuclear magnetic resonance spectroscopy are required to demonstrate the exact molecular mechanism underlying the interaction curcumin with these proteins.

This chapter summarizes evidence from *in vitro*, *in vivo*, preclinical, and clinical studies that indicated curcumin may be useful for the treatment and prevention of cancer.

2. STRUCTURE–ACTIVITY RELATIONSHIP OF CURCUMIN

Chemically, curcumin is 1,7-bis(4-hydroxy-3-methoxyphenol)-1,6-heptadiene-3,5-dione. Curcumin is made up of two methoxy-substituted aryl moieties that are linked together through a seven-carbon chain [13]. There are two ferulic acid residues and two hydrophobic phenyl domains, which are joined together by covalent bonding (Fig. 8.1). Various curcumin analogues have been reported including monoacetylcurcumin (MAC), 4-[3,5-bis[2-(4-hydroxy-3-methoxyphenyl)-ethyl]-4,5-dihydro-pyrazol-1-yl]-benzoic acid, diacetylcurcumin (DAC), difluorinated curcumin, demethoxy curcumin (DMC), isoxazolcurcumin, tetrahydrocurcumin (THC), and bisdemethoxy curcumin (BDMC) [13]. Previously, studies of curcumin structure–activity relationships showed the effects of the substitution pattern of the methoxy and hydroxy groups on the aromatic ring in curcumin. This study also suggested that removal of the hydroxy groups or inclusion of an extra

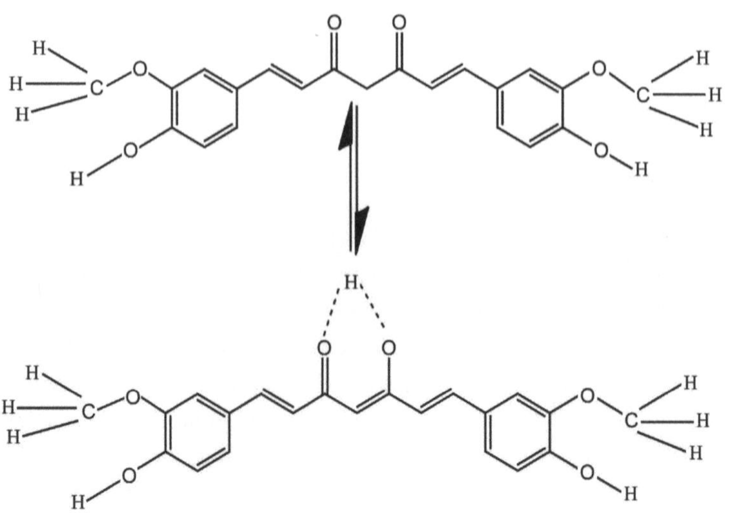

Figure 8.1 Chemical structure showing tautomerism of curcumin. Curcumin exists in keto–enol tautomer and this tautomerism favor curcumin interaction and binding with wide range of enzymes.

methoxy group on curcumin resulted in a slight decrease in its antiproliferative activity. However, no difference was observed in effect of curcumin when the methoxy groups were removed [13]. Curcumin exists in equilibrium through keto–enol tautomerism owing to its β-diketone moiety and resides entirely in the enol form both in acidic and in alkaline pH [14,15]. Studies have shown that curcuminoids are different in their physicochemical properties and pharmacological activities owing to differences in their methoxy substitutions. Methoxy substitution on the aromatic ring can optimize the interactions of curcuminoids with nucleophiles via the Michael reaction. The hydrogen bonding interaction between the phenolic hydroxy and the ortho-methoxy groups that affect oxygen–hydrogen bond energy and hydrogen atom abstraction by free radicals makes curcumin stronger antioxidant [16]. Therefore, the ortho-methoxy group and the structure–activity relationship in curcumin may be important for its anticancer activity [16,17].

3. CURCUMIN BINDS AND MODULATES MULTIPLE ENZYMES

Curcumin binds to and modulates the activity of various enzymes (Fig. 8.2), which are described below under different headings.

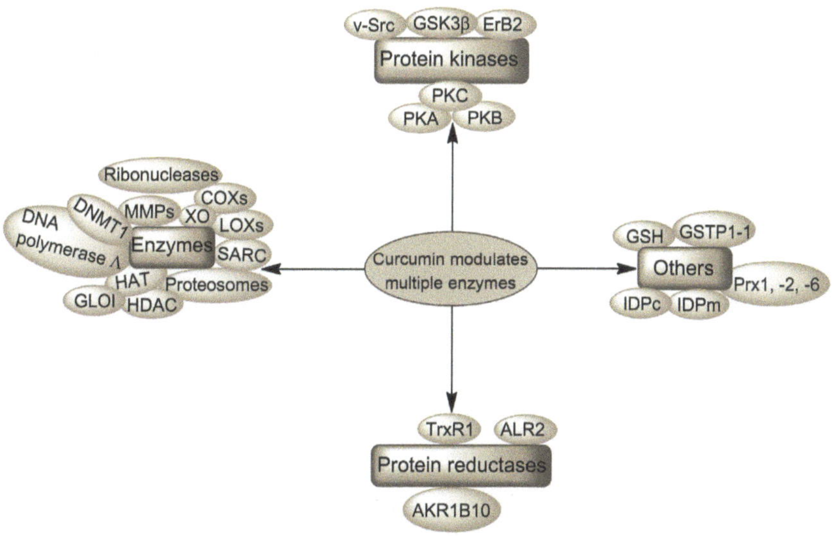

Figure 8.2 Schematic representation of multiple enzymes modulated by curcumin treatment.

3.1. Lipoxygenases

LOXs are a family of lipid-peroxidizing enzymes that catalyzes the peroxidation of arachidonic acid and implicated in the pathogenesis of various inflammatory diseases including cancers [18]. LOXs have been reported to promote cancer cell survival as well as enhance metastasis of several cancers by modulating downstream targets [18]. Therefore, use of compounds that inhibit LOX activity could be of primary importance in the treatment of cancer. Studies have shown that administration of curcumin (0.2%) reduced azoxymethane-induced colonic mucosal formation and suppressed tumor hydroxyeicosatetraenoic acid formation up to 68% in rats [19]. In addition, the inhibitory effect of curcumin on the catalytic activity of 5-LOX in HT-29 human colon cancer cells. These studies showed that curcumin efficiently inhibited the activity of human recombinant 5-LOX [20]. The underlying mechanism through which curcumin regulates the suppression of 5-LOX has not been investigated. However, it is believed that curcumin inhibited enzymatic activity of LOX by binding to the active site in a noncompetitive manner. Results from biophysical studies such as X-ray diffraction and mass spectrometry demonstrated an electron mass near the L3-LOX catalytic site [20]. It is possible that curcumin degraded and formed the peroxy complex 4-hydroperoxy-2-methoxyphenol during interaction with LOX [21]. A computational docking study with curcumin and LOX demonstrated that aromatic rings of curcumin specifically bound to LOX indicating that may be important for inhibition of LOX. Curcuminoids bond to P12-LOX effectively reduced sprout formation of endothelial cells [22]. Further studies are needed to properly address the underlying mechanism of curcumin-induced inhibition of LOX activity in various cancers.

3.2. Cyclooxygenases

The COX enzymes are known to exist in two isoforms, COX-1 and COX-2. Studies have shown that COX-1 is constitutively expressed in various tissues, whereas COX-2 is induced by growth factors, cytokines, and tumor promoters [23]. COX enzymes convert arachidonic acid to a transitional prostaglandin PGH_2, which is then converted by a specific PG synthases to PGE_2 [24]. Studies have shown that curcumin treatment inhibited lipopolysaccharide (LPS)- and phorbol-12-myristate-13-acetate (PMA)-induced COX-2 expression at the protein and mRNA levels in gastrointestinal cell lines [25]. Curcumin reduced the growth of HT-29 cells in a dose- and time-dependent fashion by blocking COX-2 specifically [26]. Curcumin also blocked ganglioside, LPS-, or interferon (IFN-γ)-induced COX-2 expression in rat primary microglial and murine BV2 microglial cells [27]. Nuclear factor-kappa B (NF-κB) and inducible nitric oxide synthase (iNOS) expression regulated COX-2 expression in various inflammatory conditions. It has been reported that curcumin downregulated COX-2 by blocking NF-κB [28,29]. Administration of curcumin suppressed azoxymethane-induced increases in both iNOS and COX-2 in colonic mucosa of colon cancer models [30–32]. The mechanism by which curcumin inhibits COX activity in inflammatory diseases has not been fully investigated. However, molecular docking studies suggested that curcumin inhibited COX-2 by direct binding. Importantly, the methoxy group of curcumin formed a covalent bond with residue Ser530 of COX-1, whereas the phenol groups of curcumin were linked by hydrogen bonding to the amino acid residues of COX-1 and COX-2 [33]. Another study strengthens the hypothesis of direct interaction of curcumin with COX-2 through hydrogen bonding with Ala562 [34]. Also, curcumin analogues (isooxazole and pyrazole) bound to COX-1 but not COX-2, indicating the potential of strong C–F bonds in pancreatic cancer cells. These studies demonstrated that COX enzymes play a critical role in various inflammatory diseases.

3.3. Xanthine Oxidase

XO is a xanthine oxidoreductase enzyme that generates reactive oxygen species, catalyzes the oxidation of hypoxanthine to xanthine, and further catalyzes the oxidation of xanthine to uric acid [35]. This enzyme has been reported to play a potential role in the progression and development of various cancers [36]. Curcumin has been reported to inhibit PMA-induced XO activation [37]. In contrast, another study reported that curcumin had no effect on XO activity [38]. Computational and molecular docking studies have shown that curcumin directly interacted with XO [39]. Curcumin bounds and interacts with XO on residues Phe914, Phe1009, and Thr1010, thus reducing their activation. Curcumin metabolites have been shown to bind XO more efficiently than the parent compound, curcumin [39]. This study highlighted the activity of the metabolites curcumin, which exhibited greater biological activity than the parent compound against XO.

3.4. Proteasomes

The proteasome is a multicatalytic proteinase complex, critical for degradation of unwanted cellular proteins by ubiquitin proteasome system (UPS).

Proteasomes may also be important targets for the effects of curcumin. Studies have shown that curcumin disrupts the UPS through suppression of the protease activities of the proteasome [40]. Curcumin inhibited the chymotrypsin-like activity of the 20S proteasome and cellular 26S proteasome by interacting with the hydroxyl group of the amino-terminal threonine residue of the proteasomal chymotrypsin-like subunit in human colon cancer HCT-116 and SW480 cells [40]. Curcumin has also been reported to be an effective inhibitor of COP9 signalosome, thus disrupting UPS [41,42]. Furthermore, curcumin inhibited the activity of COP9 signalosome-associated kinases (casein kinase 2 and protein kinase D (PKD)) and induced ubiquitination and proteasome-dependent degradation of transiently expressed Id3 in HeLa cells [41–43]. In contrast, curcumin has been shown to activate the proteasomal pathway at lower doses, whereas at higher doses inhibited this pathway [44]. However, curcumin-induced suppression of proteasome activity has been considered an alternate mechanism for the therapeutic potential of curcumin in colon cancer [40]. Furthermore, curcumin analogues such as acetates and amino acid conjugates exhibited proteasome inhibition producing a reduction in cancer cell viability. Molecular docking and computational approaches have revealed that amino acid conjugates of curcumin inhibited proteasome activity by direct binding to the β5 subunit of the proteasome [45].

3.5. Ca^{2+}-ATPase of Sarcoplasmic Reticulum

The Ca^{2+}-ATPase sarcoplasmic reticulum is the major Ca^{2+} transport protein that regulates intracellular calcium. Studies have shown that Ca^{2+}-ATPase of sarcoplasmic reticulum is inhibited by a variety of hydrophobic compounds such as curcumin. Curcumin-induced proapoptotic endoplasmic reticulum stress in human leukemia HL-60 cells by modulating apoptotic proteins such as caspase-4 and CAAT/enhancer-binding protein homologous protein [46]. Curcumin directly bound to a site within the ATPase and induced a conformational change that prevented ATP from binding, thus producing the inhibition of sarcoplasmic reticulum Ca^{2+}-ATPase in rabbit skeletal muscle [47]. In another study, curcumin-induced apoptosis and blocked the survival of human liposarcoma cells through endoplasmic reticulum stress. This inhibitory effect of curcumin is mediated through death receptor pathways as well as by direct binding with sarcoplasmic reticulum Ca^{2+}-ATPase. Molecular docking studies have shown that curcumin interacts directly through Asp254, Arg264, and Gln56 residues of the enzyme [48].

3.6. Matrix Metalloproteinases

MMPs are members of zinc-dependent endopeptidases, which are overexpressed in infiltrating tumors. MMPs are capable of degrading extracellular matrix proteins. Studies have shown that MMPs promote cell proliferation, differentiation, and angiogenesis in a variety of pathological conditions including cancers [49,50]. Among MMPs, MMP2 and MMP9 are most often involved in tumor angiogenesis [51]. It has been reported that curcumin reduced the proliferation of human breast cancer epithelial cells through suppression of MMP2 [52]. In another study, curcumin prevented the growth of brain tumors cells by downregulating the expression of MMP9 [53]. Curcumin has been shown to reduce cellular infiltration and invasion induced by the extracellular matrix protein, osteopontin. Curcumin blocked osteopontin-induced membrane-type 1 (MT1-MMP) gene expression [54]. In another study, it was reported that curcumin-mediated inhibition of 12-O-tetradecanoylphorbol-13-acetate (TPA)-induced MMP9 expression and cell invasion involves the suppression of the PKCα, mitogen-activated protein kinase, and NF-κB/activator protein pathway in MCF-7 cells [55]. Molecular docking studies revealed that the curcumin derivatives THC and BDMC interacted with MMPs through the formation of three hydrogen bonds and inhibited their activation. This study suggested that hydrogen bonds were formed between the oxygen of the curcumin heptane branch and the amine group of Arg424 [56].

3.7. Histone Acetyltransferases and Deacetylases

Both HAT and HDAC play an essential role in chromatin structure modifications via acetylation and deacetylation. Alteration and modulation of proteins by HAT such as p300/CREB-binding protein (CBP) are considered important for epigenetic regulation of gene expression. Dysregulation of these processes may contribute to carcinogenesis [57]. It has been reported that curcumin suppressed p300/CBP family of HAT proteins specifically without affecting p300/CBP-associated factor (PCAF)/GCN5 HATs. This study suggested that α-, β-unsaturated carbonyl groups in the curcumin side chain functioned as the Michael reaction sites that are required for its HAT inhibition. Curcumin has the ability to inhibit histone hyperacetylation in both PC3-M prostate cancer cells and peripheral blood lymphocytes induced by the histone deacetylase inhibitor, MS-275 [57]. Additionally, curcumin has been shown to suppress p300/CBP HAT activity-dependent chromatin transcription. Curcumin also induced apoptosis in cervical cancer cells that was associated with inhibition of acetylation of histones and p53 *in vivo* through specific inhibition of p300/CBP [58]. In another study, curcumin-induced histone hypoacetylation was reported that induced caspase-3-dependent cell death in glioma cells and neurogenesis of progenitor cells [59].

HDACs counterbalance HAT acetylation, which is critical for regulation of gene expression. Curcumin has been shown to be potent inhibitor of HDACs. Curcumin inhibited the expression of class I HDACs such as HDAC1, HDAC3, and HDAC8 and amplified the expression of Ac-histone H4 in Raji cells [60]. Recently, it was reported that curcumin inhibited HDACs such as HDAC1 and HDAC2 and caused cell cycle arrest and apoptosis by upregulating p21 expression in breast cancer MCF-7 cells [61]. In contrast, curcumin was reported to restore HDAC2 levels in the lungs of patients with chronic obstructive pulmonary disease [62]. This might be due to the multiple effects of curcumin on the subtypes of HDAC enzymes in multiple tissues. However, it has been shown that curcumin treatment induced apoptosis in medulloblastoma cells and increased survival in the Smo/Smo transgenic medulloblastoma mouse model by inhibiting HDAC4 expression [63]. Molecular docking studies showed that curcumin bound to HDAC8 and made hydrophobic contacts with active site residues of the enzyme (including Arg37, Pro35, Ile34, and Phe152) but did not react with the zinc ion located in the active site cavity [64]. Further structural conformation studies of curcumin with different subtypes of HDAC enzymes are needed to explain underlying mechanism for the effects of curcumin.

3.8. DNA Methyltransferase 1

DNA methyltransferases are a group of enzymes involved in the regulation of gene expression. DNA methyltransferases cause the aberrant DNA methylation of tumor suppressor genes, which become hypermethylated and transcriptionally silent in some blood cancers [65]. Therefore, restoration of hypermethylated tumor suppressor genes by modulation of the activity of DNA methyltransferases (DNMT1, DNMT3A, and DNMT3B) has therapeutic potential against cancer. Recently, it has been reported that curcumin inhibited the enzymatic activity of DNMT1 and prevented binding of transcription factors such as p65 and Sp1 to the DNMT1 promoter in acute myeloid leukemia (AML) cell lines. Curcumin treatment caused G1 cell cycle arrest and induced apoptosis by reactivation of the $p^{15INK4B}$ tumor suppressor gene. Curcumin administration also suppressed AML tumor growth in mice implanted with the human AML MV4–11 cells [66]. Furthermore, curcumin inhibited CpG methyltransferase (M.SssI, a DNA methyltransferase analogue) by covalently blocking the catalytic Cys1226 residue of DNMT1 in leukemia cell lines [67]. Molecular docking techniques revealed that the α-, β-unsaturated group of curcumin covalently bounds to the catalytic cysteine ofDNMT [67]. Further studies are necessary to investigate underlying mechanism of covalent binding of curcumin to DNMT1 and other DNA methyltransferases.

3.9. DNA Polymerase λ

DNA polymerase λ mediates DNA repair processes by catalyzing the polymerization of deoxyribonucleotides alongside a DNA strand. Curcumin has been reported as a potent inhibitor of polymerase λ in laboratory studies. Curcumin inhibited polymerase λ and suppressed the growth of human NUGC-3 cancer cells by mediating cell cycle arrest at the G2/M phase [68]. The same group of researchers also synthesized MAC, which bounds noncompetitively and more potently inhibited pol λ inhibitor than curcumin [69]. Computational molecular docking analyses revealed that MAC bound selectively to the N-terminal domain of pol λ, which consisted of a β-sheet (Thr51 of sheet-1), an α-helix (residues 57–69), and the two loops (residues 51–56 and 70–75) [70]. This study also demonstrated that MAC did not bind to the C-terminal region. Further studies are need to determine the binding of curcumin and curcumin derivatives with DNA polymerase λ.

3.10. Ribonucleases

RNases are group of enzymes that regulate transcription and translational cellular events and mediate alternative splicing of RNA. Aberrant expression may cause splicing defects, which can degrade RNA, and in turn inhibit protein synthesis and initiate carcinogenesis or cell death [71]. Curcumin has been shown to inhibit RNase L activity noncompetitively *in vitro*. Molecular docking studies revealed two binding sites for curcumin: one at the hydrophobic cavity near the active site and another near the junction of ankyrin repeat and kinase homology region that stabilizes the crossed homodimeric structure of RNase L [72]. Fluorescence quenching and molecular docking experiments have reported that the curcumin analogue DAC was found to interact with RNase A and modulate its activity. Molecular docking analysis demonstrated that the oxygen atoms of the keto–enol group of DAC formed hydrogen bond with the Tyr97, Gln11, and Lys7 residues of RNase, in addition to extensive nonspecific hydrophobic interactions [73]. This inhibitory effect of curcumin may be important in the therapy of pathological conditions related to upregulated RNases.

3.11. Glyoxalase I

Glyoxalases (GLO1 and GLO2) are key metalloenzyme in the glycolytic pathway involved in the detoxification of reactive methylglyoxal into D-lactate using glutathione (GSH) as cofactor. Among glyoxalases, Glo1 is an active detoxification enzyme in both cancerous and normal cells, and inhibitors of glyoxalases are regarded as potential anticancer agents [74]. Curcumin has been reported to inhibit the activity of GLO1, which causes depletion of cellular ATP and GSH. Maintenance of cellular energy balance depends on the glycolytic pathway [75]. In another study, curcumin and curcumin derivatives have been show to inhibit human GLO1. This study suggested that enol form of curcumin interacted with Zn^{2+} in the active site of GLO1 through the oxygen atoms of the carbonyl group and formed a strong hydrogen bond with Lys156, Arg122, and Arg37 residues of GLO1 [76]. In addition, the keto and enol forms of curcumin bound to the active site of GLO1, whereas the enol bound to the Glu172 and Met179 residues. The calculated binding-free energies indicated that the enol form of curcumin had greater affinity for binding to the GLO1. BDMC bound with lesser affinity than curcumin to GLO1 [77]. More studies are needed to investigate the binding affinity and interaction of curcumin with GLO2.

4. CURCUMIN BINDS AND MODULATES PKs

PKs are enzymes that catalyze phosphorylation of target proteins in signal transduction pathways. PKs have been reported to play a key role in the initiation and development of various cancers. Some of the PKs, which are modulated by curcumin action, are listed below (Fig. 8.2).

4.1. Protein Kinases

PKs are a family of serine–threonine kinases, which actively participates in cells proliferation, differentiation, transportation, and production of extracellular matrix. This family has been reported to mediate cancer initiation and development by modulating a number of downstream signaling molecules. Among the PKs, PKA, PKB, and PKC have been linked to carcinogenesis, and their inhibitors can act as tumor suppressors [78,79]. Previous studies have shown that PKA and downstream effectors regulate multiple processes, and aberrant expression of PKA has been associated with various cancers [80,81]. Curcumin has been reported to selectively inhibited PKA by noncompetitive binding [81]. Curcumin was also reported to block proliferation of breast and prostate cancer cells and induced apoptosis by direct inhibition basal activity of Akt/PKB pathway [82,83]. Curcumin suppressed TPA-induced tumor promotion through inhibition activation of PKC [84]. Molecular docking studies of a curcumin derivative with PKC C1B revealed that both molecules undergo conformational changes by forming hydrogen bonds with the tyrosine and tryptophan residues of the protein. This study also suggested that curcumin derivatives showed higher binding affinity with PKCθC1B than with PKCδC1B and PKCεC1B [85,86].

4.2. Cellular Sarcoma

Cellular sarcoma (c-Src) is a protein that is encoded by the cellular homologue of v-Src, ubiquitously expressed cytoplasmic tyrosine kinase whose overexpression has been implicated in various cancers [87]. Curcumin has been shown to inhibit Janus kinase 2 mRNA expressions as well as cyclin D1 and v-Src gene expression in K562 chronic leukemia cells [88]. In another study, curcumin directly inhibited v-Src, which resulted in the suppression of Src-mediated Shc-Tyr317 phosphorylation, focal adhesion kinase phosphorylation, decreased extracellular signal-related kinase activation, and reduced proliferation of v-Src-transformed cells [89]. Curcumin also inhibited tyrosine kinase activity of v-Src and pp60c-Src by 30% [90]. This study also confirmed that curcumin noncompetitively inhibited phosphorylase kinase (PhK), but selectively inhibited phosphorylase kinase [90].

4.3. Glycogen Synthase Kinase-3β

GSK-3β is a ubiquitously expressed protein kinase that regulates multifactorial processes through Wnt and other signaling pathways [91]. GSK-3β has been shown to phosphorylate β-catenin, which is then degraded through the UPS [92]. GSK-3β is also involved in cell cycle regulation through the phosphorylation of cyclin D1, which results in the rapid proteolytic turnover of the cyclin D1 protein. Aberrant expression of GSK-3β has been reported in various cancers [91,92]. Curcumin has been reported to inhibit GSK-3β and E-cadherin in colon cancer cell lines [93]. It is believed that curcumin fits within the binding pocket of GSK-3β through hydrogen bonding of the enolic hydroxyl group of curcumin with the amidic carbonyl of the Val135 residue. The conjugated ketone of curcumin formed hydrogen bonds with the NH group of the same amino acid residue [94].

4.4. ErbB2

ErbB2 (HER2/neu) is a transmembrane tyrosine kinase that acts as a coreceptor for the other epithelial growth factor receptor family member and activate multiple downstream pathways, some of which enhance cancer development. It has been shown that overexpression of ErbB2 increased metastasis and resistance to anticancer drugs [95]. Therefore, inhibitors that downregulate the level of ErbB2 protein and/or its activity may be potential treatments for ErbB2-overexpressing cancers. Curcumin has been reported to downregulate cellular ErbB2 in breast cancer cell lines and inhibited ErbB2-encoded p185neu tyrosine kinase [96]. Furthermore, curcumin increased the binding of the carboxyl terminus of Hsc70-interacting protein, a chaperone-dependent ubiquitin ligase with ErbB2. The binding was followed by ubiquitination and depletion of ErbB2 in SKBr3 and COS7 cells. Molecular docking studies showed that curcumin bound to the kinase domain of ErbB2, and Michael acceptor functionality was required for curcumin-induced inhibition of ErbB2 [97].

5. CURCUMIN DIRECTLY BINDS AND MODULATES PROTEIN REDUCTASES

Protein reductases are enzymes that catalyze the oxidation–reduction processes. Curcumin has the potential to modulate various protein reductases. Some of them are discussed below with regard to the activity of curcumin (Fig. 8.2).

5.1. Thioredoxin Reductase

TrxR1 catalyzes NADPH-dependent reduction of the disulfide in thioredoxin (Trx), which mediate a wide range of functions in cellular proliferation, defense against oxidative stress, apoptosis, and redox control [98]. Studies have shown that curcumin-induced inhibition of TrxR1 was required for radiosensitization of squamous carcinoma cells by curcumin [99]. In another study, curcumin irreversibly inhibited TrxR1 by covalent binding of curcumin to the Cys496/Sec497 residues in the active site of TrxR1 through Michael addition [100]. Additionally, molecular docking analysis was performed to compare the TrxR inhibitory effects of curcumin, DMC, and BDMC. This study suggested that curcumin mainly interacted with residues His108, Arg351, Lys29, and Leu112 in the F chain of TrxR. In addition, the Se atom of the SeCys498 residue and formed a hydrophobic contact with the sp2 carbon of the β-diketone moiety of curcumin. These studies confirmed that the methoxy group in curcuminoids was essential for interaction with catalytic residues of Trx [101]. Further studies are needed to demonstrate curcumin and TrxR interaction and underlying mechanisms of TrxR inhibition by curcumin.

5.2. Aldose Reductase

AR is important member of aldo–keto reductase super family. AR reduces glucose to sorbitol utilizing NADPH as a cofactor and regulates reactive oxygen species induced by growth factors and cytokines [102]. It has been reported that inhibition or silencing of AR could prevent growth factor-induced activation of inflammatory markers such as NF-κB, expression of COX-2, and formation of PGE_2 in colon cancer cells [102]. Therefore, compounds that inhibit AR may be important in the prevention and treatment of various cancers. Curcumin selectively inhibited ALR2 activity in a noncompetitive manner. Molecular docking studies demonstrated that curcumin interacts with ALR2 at residues Tyr48, Lys21, Thr19, Gln183, Leu300, and Trp111 in the active site [103]. Furthermore, human small intestine reductase or AKR1B10 is another member of aldo–keto reductase family, which is overexpressed in a number of cancer cells, and AKR1B10 was involved in cell survival and protection against anticancer drugs [104,105]. Like ALR2, curcumin inhibited AKR1B10 with IC50 of 30 μM [103]. In addition, BDMC (curcumin analogue) was found to be a more potent inhibitor of AKR1B10 than curcumin. Molecular docking analysis demonstrated that the Trp21, Gln114, Trp220, Val301, and Ser304 residues of AKR1B10 formed hydrogen bonds with the hydroxyl group on the phenyl ring of BDMC [104,105].

6. OTHERS

NADPH is essential for the reconstitution of the reduced GSH pool, thereby providing protection to cells against oxidative damage [106,107]. In addition, ICDH including $NADP^+$-dependent IDPm and IDPc and thiol-specific peroxidases mediate antioxidant function during oxidative stress. In addition to their antioxidant role, these enzymes also participate in various biological processes such as cell proliferation and apoptosis [106,107]. Recent studies have shown that curcumin inhibited the activation of IDPm, IDPc, Prx1, Prx2, and Prx6, and thereby radiosensitized colon cancer cells and reduced tumor growth of B16F10-induced tumors in C57BL/6J mice [108]. Previously, a complex interaction between curcumin and GSH has been reported [109]. This specific interaction between curcumin with GSH consequently modulated recombinant human glutathione *S*-transferase (GSTP1-1). Biophysical analysis revealed that the presence of GSTP1-1 considerably enhanced GSH-mediated binding of curcumin [109]. Curcumin analogs EF24 and EF31 have been reported to mediate anticancer effects by modulation of GSH and cysteine-containing dipeptides in breast cancer cells [110].

7. CURCUMIN CLINICAL TRIALS IN CANCER

Curcumin has the ability to inhibit signaling pathways and molecular targets involved in the initiation and progression of various cancers. To date, results from 66 clinical trials have been published, and 35 clinical trials are still in progress (https://clinicaltrials.gov). Curcumin (36–180 mg/day) was given to colon cancer patients for up to 4 months. A decline in levels of the cancer biomarker carcinoembryonic antigen from 310 ± 15 to 175 ± 9 ng/ml was observed after 2 months of curcumin administration [111]. Curcumin capsules containing three different doses (450, 1800, or 3600 mg/day) were taken by 12 patients for 7 days. Curcumin treatment decreased the levels of pyrimidopurinone-deoxyguanosine M1G adducts in malignant colorectal tissue, but the levels of COX-2 were unaffected by the same dose of curcumin [112]. A nonrandomized, open-label clinical trial was conducted in 44 confirmed smokers in which curcumin was given orally at two different doses (2 or 4 g/day) for 30 days. Curcumin treatment reduced the formation of aberrant crypt foci, the precursors to colorectal polyps [113]. Furthermore, curcumin capsules (360 mg thrice/day) improved the general health of colorectal cancer patients via increased p53 expression after diagnosis and presurgery [114].

Curcumin (500 mg) and piperine (5 mg) were given daily to 20 patients with tropical pancreatitis, which resulted in a reduction in erythrocyte malondialdehyde levels with a significant increase in GSH levels. Oral administration of curcumin (8 g) to pancreatic cancer patients decreased the expression of COX-2 [115,116]. In another clinical trial, curcumin (8 g) was orally administered daily to 25 patients for 2 months. In this study, one patient showed marked tumor regression accompanied by significant increase in levels of serum cytokines (IL-6, IL-8, IL-10, and IL-1). These findings demonstrate that orally administered curcumin was well tolerated and has biological activity in some patients with pancreatic cancer [117].

Zyflamend (curcumin) was given to 23 patients of high-grade prostatic intraepithelial neoplasia (HGPIN) three times a day for 18 months. A biopsy conducted at 18 months revealed no markers of HGPIN and a reduction in NF-κB and C-reactive protein [118]. In a phase I clinical trial, a daily 0.5–12 g oral dose of curcumin administered for 3 months resulted in the histological improvement of precancerous lesions in one of four patients suffering from cervical intraepithelial neoplasia [119]. In a randomized, double-blind, placebo-controlled crossover study, 1 and 8 g of curcumin were given to monoclonal gammopathy of undetermined significance and smoldering multiple myeloma patients [120,121]. Curcumin administration decreased the free light chain ratio, urinary deoxypyridinoline, and serum creatinine levels compared to the placebo group [121].

The efficacy of curcumin was also investigated in patients with squamous cell carcinoma of the head and neck (HNSCC). Curcumin treatment reduced IKKβ kinase activity in the salivary cells and improved dental caries in patients with HNSCC [122]. The effect of a 1 g curcumin tablet (900 mg of curcumin, 80 mg of DMC, and 20 mg of BDMC) was investigated in patients with oral leukoplakia, oral submucous fibrosis, or lichen planus and healthy individuals. Curcumin treatment decreased malondialdehyde and 8-hydroxydeoxyguanosine and increased vitamins C and E in the serum and saliva of patients with precancerous lesions [123]. In another clinical trials, the safety of curcuminoids (6 g/day) was investigated in a randomized, double-blind, placebo-controlled clinical trial of 20 patients with oral lichen planus. This study showed that curcuminoids at 6 g/day are well tolerated and may prove efficacious in controlling signs and symptoms of oral lichen planus [124].

8. FUTURE PERSPECTIVES

The studies described above clearly indicate that curcumin treatment suppressed initiation and development of various cancers. Curcumin modulated multiple enzymes, and this all combined to advocate pleiotropic effects of this dietary supplement. The chemical structure of curcumin and its analogues enabled their physicochemical interaction with molecular targets in various cancers. The aromatic groups provided hydrophobicity, and the methoxy substitution enabled curcumin to bind and modulate a variety of enzymes. These multiple molecular interactions of curcumin with the target enzymes make it a potentially effective chemotherapeutic agent for cancer. Due to the complexity of cancer, the underlying mechanism of curcumin-induced modulation of many enzymes in some cancers remains unclear.

A number of biophysical methods have been adopted to study the interactions of curcumin with enzymes. Among these methods, molecular docking was the most valuable method for studying the interaction and binding affinity of curcumin with target enzymes. However, more biophysical studies are needed to elaborate the exact underlying molecular mechanisms of curcumin interaction with enzymes.

There is compelling evidence that curcumin exerts its anticancer effects by modulating multiple molecular targets. However, these effects have not been translated to the clinic for the treatment of cancer, even though curcumin has been found to be safe and well tolerated. Therefore, future studies should utilize biophysical technologies to identify new molecular targets of curcumin. In addition, large scale pharmacodynamic studies should be conducted to determine adverse effects. At present, there is not sufficient data to support the recommendation for the long-term use of curcumin for the prevention or treatment of cancer. Future computational and clinical studies are urgently needed to uncover the underlying mechanisms of action for curcumin in various cancers.

ACKNOWLEDGMENT

This study was supported by Basic Science Research Program through the NRF funded by the Ministry of Science, ICT & Future Planning (NRF-2013R1A1A2063612).

REFERENCES

[1] A. Shehzad, F. Wahid, Y.S. Lee, Curcumin in cancer chemoprevention: Molecular targets, pharmacokinetics, bioavailability, and clinical trials, Arch. Pharm. (Weinheim) 343 (2010) 489–499.
[2] A. Shehzad, Y.S. Lee, Molecular mechanisms of curcumin action: Signal transduction, Biofactors 39 (2013) 27–36.
[3] A. Shehzad, J. Lee, Y.S. Lee, Curcumin in various cancers, Biofactors 39 (2013) 56–68.
[4] WHO, Evaluation of Certain Food Additives, WHO Technical Report Series, 891, WHO, Geneva, 2000.

[5] C.D. Lao, M.T. Ruffin, D. Normolle, D.D. Heath, S.I. Murray, J.M. Bailey, M.E. Boggs, J. Crowell, C.L. Rock, D.E. Brenner, Dose escalation of a curcuminoid formulation, BMC Complement. Altern. Med. 6 (2006) 10.
[6] R. Siegel, J. Ma, Z. Zou, A. Jemal, Cancer statistics, 2014, CA Cancer J. Clin. 64 (2014) 9–29.
[7] M. González-Vallinas, M. González-Castejón, A. Rodríguez-Casado, A. Ramírez de Molina, Dietary phytochemicals in cancer prevention and therapy: A complementary approach with promising perspectives, Nutr. Rev. 71 (2013) 585–599.
[8] S.C. Gupta, S. Prasad, J.H. Kim, S. Patchva, L.J. Webb, I.K. Priyadarsini, B.B. Aggarwal, Multitargeting by curcumin as revealed by molecular interaction studies, Nat. Prod. Rep. 28 (2011) 1937–1955.
[9] A. Mukhopadhyay, S. Banerjee, L.J. Stafford, C. Xia, M. Liu, B.B. Aggarwal, Curcumin-induced suppression of cell proliferation correlates with down-regulation of cyclin D1 expression and CDK4-mediated retinoblastoma protein phosphorylation, Oncogene 21 (2002) 8852–8861.
[10] R. Garg, S. Gupta, G.B. Maru, Dietary curcumin modulates transcriptional regulators of phase I and phase II enzymes in benzo[a]pyrene-treated mice: Mechanism of its anti-initiating action, Carcinogenesis 29 (2008) 1022–1032.
[11] S. Shishodia, P. Potdar, C.G. Gairola, B.B. Aggarwal, Curcumin (diferuloylmethane) down-regulates cigarette smoke-induced NF-kappaB activation through inhibition of IkappaBalpha kinase in human lung epithelial cells: Correlation with suppression of COX-2, MMP-9 and cyclin D1, Carcinogenesis 24 (2003) 1269–1279.
[12] G. Xu, Y. Chu, N. Jiang, J. Yang, F. Li, The three dimensional quantitative structure activity relationships (3D-QSAR) and docking studies of curcumin derivatives as androgen receptor antagonists, Int. J. Mol. Sci. 13 (2012) 6138–6155.
[13] J.R. Fuchs, B. Pandit, D. Bhasin, J.P. Etter, N. Regan, D. Abdelhamid, C. Li, J. Lin, P.K. Li, Structure-activity relationship studies of curcumin analogues, Bioorg. Med. Chem. Lett. 19 (2009) 2065–2069.
[14] X. Lin, S. Ji, R. Li, Y. Dong, X. Qiao, H. Hu, W. Yang, D. Guo, P. Tu, M. Ye, Terpecurcumins A-I from the rhizomes of *Curcuma longa*: Absolute configuration and cytotoxic activity, J. Nat. Prod. 75 (2012) 2121–2131.
[15] T. Masuda, T. Maekawa, K. Hidaka, H. Bando, Y. Takeda, H. Yamaguchi, Chemical studies on antioxidant mechanism of curcumin: Analysis of oxidative coupling products from curcumin and linoleate, J. Agric. Food Chem. 49 (2001) 2539–2547.
[16] P. Somparn, C. Phisalaphong, S. Nakornchai, S. Unchern, N.P. Morales, Comparative antioxidant activities of curcumin and its demethoxy and hydrogenated derivatives, Biol. Pharm. Bull. 30 (2007) 74–78.
[17] D.F. Birt, J. Przybyszewski, W. Wang, J. Stewart, Y. Liu, Identification of molecular targets for dietary energy restriction prevention of skin carcinogenesis. An idea cultivated by Edward Bresnick, J. Cell. Biochem. 91 (2004) 258–264.
[18] A. Comba, M.E. Pasqualini, Primers on molecular pathways-lipoxygenases: Their role as an oncogenic pathway in pancreatic cancer, Pancreatology 9 (2009) 724–728.
[19] C.V. Rao, A. Rivenson, B. Simi, B.S. Reddy, Chemoprevention of colon carcinogenesis by dietary curcumin, a naturally occurring plant phenolic compound, Cancer Res. 55 (1995) 259–266.
[20] J. Hong, M. Bose, J. Ju, J.H. Ryu, X. Chen, S. Sang, M.J. Lee, C.S. Yang, Modulation of arachidonic acid metabolism by curcumin and related beta-diketone derivatives; effects of cytosolic phospholipase A2, cyclooxygenases and 5-lipoxygenase, Carcinogenesis 25 (2004) 1671–1679.
[21] E. Skrzypczak-Jankun, K. Zhou, N.P. McCabe, S.H. Selman, J. Jankun, Structure of curcumin in complex with lipoxygenase and its significance in cancer, Int. J. Mol. Med. 12 (2003) 17–24.

[22] J. Jankun, A.M. Aleem, S. Malgorzewicz, M. Szkudlarek, M.I. Zavodszky, D.L. Dewitt, M. Feig, S.H. Selman, E. Skrzypczak-Jankun, Synthetic curcuminoids modulate the arachidonic acid metabolism of human platelet 12-lipoxygenase and reduce sprout formation of human endothelial cells, Mol. Cancer Ther. 5 (2006) 1371–1382.

[23] R.T. Poon, I.O. Ng, C. Lau, L.X. Zhu, W.C. Yu, C.M. Lo, S.T. Fan, J. Wong, Serum vascular endothelial growth factor predicts venous invasion in hepatocellular carcinoma: A prospective study, Ann. Surg. 233 (2001) 227–235.

[24] I. Venza, L. Giordano, G. Piraino, N. Medici, G. Ceci, D. Teti, Prostaglandin E2 signalling pathway in human T lymphocytes from healthy and conjunctiva basal cell carcinoma-bearing subjects, Immunol. Cell Biol. 79 (2001) 482–489.

[25] F. Zhang, N.K. Altorki, J.R. Mestre, K. Subbaramaiah, A.J. Dannenberg, Inhibition of cyclo-oxygenase 2 expression in colon cells by the chemopreventive agent curcumin involves inhibition of NF-kappaB activation via the NIK/IKK signaling complex, Carcinogenesis 20 (1999) 445–451.

[26] A. Goel, C.R. Boland, D.P. Chauhan, Specific inhibition of cyclooxygenase-2 (COX-2) expression by dietary curcumin in HT-29 human colon cancer cells, Cancer Lett. 172 (2001) 111–118.

[27] H.Y. Kim, E.J. Park, E.H. Joe, I. Jou, Curcumin suppresses Janus kinase-STAT inflammatory signaling through activation of Src homology 2 domain-containing tyrosine phosphatase 2 in brain microglia, J. Immunol. 171 (2003) 6072–6079.

[28] Y.J. Surh, K.S. Chun, H.H. Cha, S.S. Han, Y.S. Keum, K.K. Park, S.S. Lee, Molecular mechanism underlying chemopreventive activities of anti-inflammatory phytochemicals: Down regulation of COX-2 and iNOS through suppression of NF-kappa B activation, Mutat. Res. 480 (2001) 243–268.

[29] A.A. Nanji, K. Jokelainen, G.L. Tipoe, A. Rahemtulla, P. Thomas, A.J. Dannenberg, Curcumin prevents alcohol-induced liver disease in rats by inhibiting the expression of NF-kappa B-dependent genes, Am. J. Physiol. Gastrointest. Liver Physiol. 284 (2003) 321–327.

[30] C.V. Rao, C. Indranie, B. Simi, P.T. Manning, J.R. Connor, B.S. Reddy, Chemopreventive properties of a selective inducible nitric oxide synthase inhibitor in colon carcinogenesis, administered alone or in combination with celecoxib, a selective cyclooxygenase inhibitor, Cancer Res. 62 (2002) 165–170.

[31] C.V. Rao, T. Kawamori, R. Hamid, B.S. Reddy, Chemoprevention of colonic aberrant crypt foci by an inducible nitric oxide synthase-selective inhibitor, Carcinogenesis 20 (1999) 641–644.

[32] C.V. Rao, I. Cooma, M.V. Swamy, B. Simi, B.S. Reddy, Modulation of inducible nitric oxide synthase and cyclooxygenase activities by curcumin during different stages of experimental colon carcinogenesis, Proc. Am. Assoc. Cancer Res. 39 (2001) 3084.

[33] C. Selvam, S.M. Jachak, R. Thilagavathi, A.K. Chakraborti, Design, synthesis, biological evaluation and molecular docking of curcumin analogues as antioxidant, cyclooxygenase inhibitory and anti-inflammatory agents, Bioorg. Med. Chem. Lett. 15 (2005) 1793–1797.

[34] S. Padhye, S. Banerjee, D. Chavan, S. Pandye, K.V. Swamy, S. Ali, J. Li, Q.P. Dou, F.H. Sarkar, Fluorocurcumins as cyclooxygenase-2 inhibitor: Molecular docking, pharmacokinetics and tissue distribution in mice, Pharm. Res. 26 (2009) 2438–2445.

[35] C. Griot, M. Vandevelde, A. Richard, E. Peterhans, R. Stocker, Selective degeneration of oligodendrocytes mediated by reactive oxygen species, Free Radic. Res. Commun. 11 (1990) 181–193.

[36] R. Harrison, Structure and function of xanthine oxidoreductase: Where are we now? Free Radic. Biol. Med. 33 (2002) 774–797.

[37] J.K. Lin, C.A. Shih, Inhibitory effect of curcumin on xanthine dehydrogenase/oxidase induced by phorbol-12-myristate-13-acetate in NIH3T3 cells, Carcinogenesis 15 (1994) 1717–1721.

[38] J.M. Pauff, R. Hille, Inhibition studies of bovine xanthine oxidase by luteolin, silibinin, quercetin, and curcumin, J. Nat. Prod. 72 (2009) 725–731.
[39] L. Shen, H.F. Ji, Insights into the inhibition of xanthine oxidase by curcumin, Bioorg. Med. Chem. Lett. 19 (2009) 5990–5993.
[40] V. Milacic, S. Banerjee, K.R. Landis-Piwowar, F.H. Sarkar, A.P. Majumdar, Q.P. Dou, Curcumin inhibits the proteasome activity in human colon cancer cells in vitro and in vivo, Cancer Res. 68 (2008) 7283–7292.
[41] W. Henke, K. Ferrell, D. Bech-Otschir, M. Seeger, R. Schade, P. Jungblut, M. Naumann, W. Dubiel, Comparison of human COP9 signalsome and 26S proteasome lid, Mol. Biol. Rep. 26 (1999) 29–34.
[42] S. Uhle, O. Medalia, R. Waldron, R. Dumdey, P. Henklein, D. Bech-Otschir, X. Huang, M. Berse, J. Sperling, R. Schade, Protein kinase CK2 and protein kinase D are associated with the COP9 signalosome, EMBO J. 22 (2003) 1302–1312.
[43] M. Berse, M. Bounpheng, X. Huang, B. Christy, C. Pollmann, W. Dubiel, Ubiquitin-dependent degradation of Id1 and Id3 is mediated by the COP9 signalosome, J. Mol. Biol. 343 (2004) 361–370.
[44] R.E. Ali, S.I. Rattan, Curcumin's biphasic hormetic response on proteasome activity and heat-shock protein synthesis in human keratinocytes, Ann. N. Y. Acad. Sci. 1067 (2006) 394–399.
[45] S.B. Wan, H. Yang, Z. Zhou, Q.C. Cui, D. Chen, J. Kanwar, I. Mohammad, Q.P. Dou, T.H. Chan, Evaluation of curcumin acetates and amino acid conjugates as proteasome inhibitors, Int. J. Mol. Med. 26 (2010) 447–455.
[46] H.O. Pae, S.O. Jeong, G.S. Jeong, K.M. Kim, H.S. Kim, S.A. Kim, Y.C. Kim, S.D. Kang, B.N. Kim, H.T. Chung, Curcumin induces pro-apoptotic endoplasmic reticulum stress in human leukemia HL-60 cells, Biochem. Biophys. Res. Commun. 353 (2007) 1040–1045.
[47] J.G. Bilmen, S.Z. Khan, M.H. Javed, F. Michelangeli, Inhibition of the SERCA Ca2+ pumps by curcumin. Curcumin putatively stabilizes the interaction between the nucleotide-binding and phosphorylation domains in the absence of ATP, Eur. J. Biochem. 268 (2001) 6318–6327.
[48] L. Wang, L. Wang, R. Song, Y. Shen, Y. Sun, Y. Gu, Y. Shu, Q. Xu, Targeting sarcoplasmic/endoplasmic reticulum Ca^{2+}-ATPase 2 by curcumin induces ER stress-associated apoptosis for treating human liposarcoma, Mol. Cancer Ther. 10 (2011) 461–471.
[49] M. Egeblad, Z. Werb, New functions for the matrix metalloproteinases in cancer progression, Nat. Rev. Cancer 2 (2002) 161–174.
[50] W.C. Parks, C.L. Wilson, Y.S. López-Boado, Matrix metalloproteinases as modulators of inflammation and innate immunity, Nat. Rev. Immunol. 4 (2004) 617–629.
[51] S. Zucker, J. Vacirca, Role of matrix metalloproteinases (MMPs) in colorectal cancer, Cancer Metastasis Rev. 23 (2004) 101–117.
[52] Y.R. Chen, T.H. Tan, Inhibition of the c-Jun N-terminal kinase (JNK) signaling pathway by curcumin, Oncogene 17 (1998) 173–178.
[53] M.S. Woo, S.H. Jung, S.Y. Kim, J.W. Hyun, K.H. Ko, W.K. Kim, H.S. Kim, Curcumin suppresses phorbol ester-induced matrix metalloproteinase-9 expression by inhibiting the PKC to MAPK signaling pathways in human astroglioma cells, Biochem. Biophys. Res. Commun. 335 (2005) 1017–1025.
[54] S. Philip, A. Bulbule, G.C. Kundu, Matrix metalloproteinase-2: Mechanism and regulation of NF-kappaB-mediated activation and its role in cell motility and ECM-invasion, Glycoconj. J. 21 (2004) 429–441.
[55] J.M. Kim, E.M. Noh, K.B. Kwon, J.S. Kim, Y.O. You, J.K. Hwang, et al., Curcumin suppresses the TPA-induced invasion through inhibition of PKCα-dependent MMP-expression in MCF-7 human breast cancer cells, Phytomedicine 19 (2012) 1085–1092.

[56] C.R. Girija, K. Prasantha, S.P. Chtan, S.B. Noor, A.S. Akheel, Molecular docking studies of curcumin derivatives with multiple protein targets for procarcinogen activating enzyme inhibition, J. Proteomics Bioinform. 3 (2010) 200–203.
[57] M.G. Marcu, Y.J. Jung, S. Lee, E.J. Chung, M.J. Lee, J. Trepel, L. Neckers, Curcumin is an inhibitor of p300 histone acetylatransferase, Med. Chem. 2 (2006) 169–174.
[58] K. Balasubramanyam, R.A. Varier, M. Altaf, V. Swaminathan, N.B. Siddappa, U. Ranga, T.K. Kundu, Curcumin, a novel p300/CREB-binding protein-specific inhibitor of acetyltransferase, represses the acetylation of histone/nonhistone proteins and histone acetyltransferase-dependent chromatin transcription, J. Biol. Chem. 279 (2004) 51163–51171.
[59] S.K. Kang, S.H. Cha, H.G. Jeon, Curcumin-induced histone hypoacetylation enhances caspase-3-dependent glioma cell death and neurogenesis of neural progenitor cells, Stem Cells Dev. 15 (2006) 165–174.
[60] H.L. Liu, Y. Chen, G.H. Cui, J.F. Zhou, Curcumin, a potent anti-tumor reagent, is a novel histone deacetylase inhibitor regulating B-NHL cell line Raji proliferation, Acta Pharmacol. Sin. 26 (2005) 603–609.
[61] S. Mukherjeea, R. Sarkara, J. Biswasb, M. Roya, Curcumin inhibits histone deacetylase leading to cell cycle arrest and apoptosis via upregulation of p21 in breast cancer cell lines, Int. J. Green Nanotechnol. 4 (2012) 183–197.
[62] K.K. Meja, S. Rajendrasozhan, D. Adenuga, S.K. Biswas, I.K. Sundar, G. Spooner, et al., Curcumin restores corticosteroid function in monocytes exposed to oxidants by maintaining HDAC2, Am. J. Respir. Cell Mol. Biol. 39 (2008) 312–323.
[63] S.J. Lee, C. Krauthauser, V. Maduskuie, P.T. Fawcett, J.M. Olson, S.A. Rajasekaran, Curcumin-induced HDAC inhibition and attenuation of medulloblastoma growth in vitro and in vivo, BMC Cancer 11 (2011) 144.
[64] G. Bora-Tatar, D. Dayangaç-Erden, A.S. Demir, S. Dalkara, K. Yelekçi, H. Erdem-Yurter, Molecular modifications on carboxylic acid derivatives as potent histone deacetylase inhibitors: Activity and docking studies, Bioorg. Med. Chem. 17 (2009) 5219–5228.
[65] J.G. Herman, S.B. Baylin, Gene silencing in cancer in association with promoter hypermethylation, N. Engl. J. Med. 349 (2003) 2042–2054.
[66] J. Yu, Y. Peng, L.C. Wu, Z. Xie, Y. Deng, T. Hughes, et al., Curcumin down-regulates DNA methyltransferase 1 and plays an anti-leukemic role in acute myeloid leukemia, PLoS One 8 (2013) e55934.
[67] Z. Liu, Z. Xie, W. Jones, R.E. Pavlovicz, S. Liu, J. Yu, et al., Curcumin is a potent DNA hypomethylation agent, Bioorg. Med. Chem. Lett. 19 (2008) 706–709.
[68] Y. Mizushina, M. Hirota, C. Murakami, T. Ishidoh, S. Kamisuki, N. Shimazaki, et al., Some anti-chronic inflammatory compounds are DNA polymerase lambda-specific inhibitors, Biochem. Pharmacol. 66 (2003) 1935–1944.
[69] Y. Mizushina, T. Ishidoh, T. Takeuchi, N. Shimazaki, O. Koiwai, K. Kuramochi, S. Kobayashi, F. Sugawara, K. Sakaguchi, H. Yoshida, Monoacetylcurcumin: A new inhibitor of eukaryotic DNA polymerase lambda and a new ligand for inhibitor-affinity chromatography, Biochem. Biophys. Res. Commun. 337 (2005) 1288–1295.
[70] T. Takeuchi, T. Ishidoh, H. Iijima, I. Kuriyama, N. Shimazaki, O. Koiwai, Structural relationship of curcumin derivatives binding to the BRCT domain of human DNA polymerase lambda, Genes Cells 11 (2006) 223–235.
[71] S. Chen, Y. Xue, X. Wu, C. Le, A. Bhutkar, E.L. Bell, F. Zhang, R. Langer, P.A. Sharp, Global microRNA depletion suppresses tumor angiogenesis, Genes Dev. 28 (2014) 1054–1067.
[72] A. Gupta, P.C. Rath, Curcumin, a natural antioxidant, acts as a noncompetitive inhibitor of human RNase L in presence of its cofactor 2-5A in vitro, Biomed Res. Int. 2014 (2014) 1–9.

[73] B.K. Sahoo, K.S. Ghosh, S. Dasgupta, An investigation of the molecular interactions of diacetylcurcumin with ribonuclease A, Protein Pept. Lett. 16 (2009) 1485–1495.
[74] P.K. Padmanabhan, A. Mukherjee, R. Madhubala, Characterization of the gene encoding glyoxalase II from Leishmania donovani: A potential target for anti-parasite drugs, Biochem. J. 393 (2006) 227–234.
[75] T. Santel, G. Pflug, N.Y. Hemdan, A. Schäfer, M. Hollenbach, M. Buchold, et al., Curcumin inhibits glyoxalase 1: A possible link to its anti-inflammatory and anti-tumor activity, PLoS One 3 (2008) e3508.
[76] M. Yuan, M. Luo, Y. Song, Q. Xu, X. Wang, Y. Cao, X. Bu, Y. Ren, X. Hu, Identification of curcumin derivatives as human glyoxalase I inhibitors: A combination of biological evaluation, molecular docking, 3D-QSAR and molecular dynamics simulation studies, Bioorg. Med. Chem. 19 (2011) 1189–1196.
[77] M. Liu, M. Yuan, M. Luo, X. Bu, H.B. Luo, X. Hu, Binding of curcumin with glyoxalase I: Molecular docking, molecular dynamics simulations, and kinetics analysis, Biophys. Chem. 147 (2010) 28–34.
[78] E.M. Griner, M.G. Kazanietz, Protein kinase C and other diacylglycerol effectors in cancer, Nat. Rev. Cancer 7 (2007) 281–294.
[79] S. Knapp, M. Sundström, Recently targeted kinases and their inhibitors-the path to clinical trials, Curr. Opin. Pharmacol. 17 (2014) 58–63.
[80] H.S. Yeo, A. Shehzad, Y.S. Lee, Prostaglandin E2 blocks menadione-induced apoptosis through the Ras/Raf/Erk signaling pathway in promonocytic leukemia cell lines, Mol. Cells 33 (2012) 371–378.
[81] S. Reddy, B.B. Aggarwal, Curcumin is a non-competitive and selective inhibitor of phosphorylase kinase, FEBS Lett. 341 (1994) 19–22.
[82] M.S. Squires, E.A. Hudson, L. Howells, S. Sale, C.E. Houghton, J.L. Jones, et al., Relevance of mitogen activated protein kinase (MAPK) and phosphotidylinositol-3-kinase/protein kinase B (PI3K/PKB) pathways to induction of apoptosis by curcumin in breast cells, Biochem. Pharmacol. 65 (2003) 361–376.
[83] L.R. Chaudhary, K.A. Hruska, Inhibition of cell survival signal protein kinase B/Akt by curcumin in human prostate cancer cells, J. Cell. Biochem. 89 (2003) 1–5.
[84] J.Y. Liu, S.J. Lin, J.K. Lin, Inhibitory effects of curcumin on protein kinase C activity induced by 12-O-tetradecanoyl-phorbol-13-acetate in NIH 3T3 cells, Carcinogenesis 14 (1993) 857–861.
[85] A. Majhi, G.M. Rahman, S. Panchal, J. Das, Binding of curcumin and its long chain derivatives to the activator binding domain of novel protein kinase C, Bioorg. Med. Chem. 18 (2010) 1591–1598.
[86] J. Das, S. Pany, S. Panchal, A. Majhi, G.M. Rahman, Binding of isoxazole and pyrazole derivatives of curcumin with the activator binding domain of novel protein kinase C, Bioorg. Med. Chem. 19 (2011) 6196–6202.
[87] M.K. White, J.S. Pagano, K. Khalili, Viruses and human cancers: A long road of discovery of molecular paradigms, Clin. Microbiol. Rev. 27 (2014) 463–481.
[88] R. Blasius, S. Reuter, E. Henry, M. Dicato, M. Diederich, Curcumin regulates signal transducer and activator of transcription (STAT) expression in K562 cells, Biochem. Pharmacol. 72 (2006) 1547–1554.
[89] T.H. Leu, S.L. Su, Y.C. Chuang, M.C. Maa, Direct inhibitory effect of curcumin on Src and focal adhesion kinase activity, Biochem. Pharmacol. 66 (2003) 2323–2331.
[90] E.P. Moiseeva, E.P. Moiseeva, Effect of dietary phytochemicals on Src EGFR and HER2 in cancer cells, in: C.C. Earl (Ed.), Oncogene Proteins, New Research, Nova Science Publishers, New York, 2008, pp. 65–98.

[91] J. Luo, Glycogen synthase kinase 3beta (GSK3beta) in tumorigenesis and cancer chemotherapy, Cancer Lett. 273 (2009) 194–200.

[92] C. Liu, Y. Li, M. Semenov, C. Han, G.H. Baeg, Y. Tan, et al., Control of beta-catenin phosphorylation/degradation by a dual-kinase mechanism, Cell 108 (2002) 837–847.

[93] A.S. Jaiswal, B.P. Marlow, N. Gupta, S. Narayan, Beta-catenin-mediated transactivation and cell-cell adhesion pathways are important in curcumin (diferuylmethane)-induced growth arrest and apoptosis in colon cancer cells, Oncogene 21 (2002) 8414–8427.

[94] Y. Bustanji, M.O. Taha, I.M. Almasri, M.A. Al-Ghussein, M.K. Mohammad, H.S. Alkhatib, Inhibition of glycogen synthase kinase by curcumin: Investigation by simulated molecular docking and subsequent in vitro/in vivo evaluation, J. Enzyme Inhib. Med. Chem. 24 (2009) 771–778.

[95] N.G. Anderson, T. Ahmad, ErbB receptor tyrosine kinase inhibitors as therapeutic agents, Front. Biosci. 7 (2002) 1926–1940.

[96] R.L. Hong, W.H. Spohn, M.C. Hung, Curcumin inhibits tyrosine kinase activity of p185neu and also depletes p185neu, Clin. Cancer Res. 5 (1999) 1884–1891.

[97] Y. Jung, W. Xu, H. Kim, N. Ha, L. Neckers, Curcumin-induced degradation of ErbB2: A role for the E3 ubiquitin ligase CHIP and the Michael reaction acceptor activity of curcumin, Biochim. Biophys. Acta 1773 (2007) 383–390.

[98] A. Mollbrink, R. Jawad, A. Vlamis-Gardikas, P. Edenvik, B. Isaksson, O. Danielsson, et al., Expression of thioredoxins and glutaredoxins in human hepatocellular carcinoma: Correlation to cell proliferation, tumor size and metabolic syndrome, Int. J. Immunopathol. Pharmacol. 27 (2014) 169–183.

[99] P. Javvadi, L. Hertan, R. Kosoff, T. Datta, J. Kolev, R. Mick, et al., Thioredoxin reductase-1 mediates curcumin-induced radiosensitization of squamous carcinoma cells, Cancer Res. 70 (2010) 1941–1950.

[100] J. Fang, J. Lu, A. Holmgren, Thioredoxin reductase is irreversibly modified by curcumin: A novel molecular mechanism for its anticancer activity, J. Biol. Chem. 280 (2005) 25284–25290.

[101] D.V. Singh, K. Misra, Curcuminoids as inhibitors of thioredoxin reductase: A receptor based pharmacophore study with distance mapping of the active site, Bioinformation 4 (2009) 187–192.

[102] R. Tammali, A.B. Reddy, A. Saxena, P.G. Rychahou, B.M. Evers, S. Qiu, et al., Inhibition of aldose reductase prevents colon cancer metastasis, Carcinogenesis 32 (2011) 1259–1267.

[103] P. Muthenna, P. Suryanarayana, S.K. Gunda, J.M. Petrash, G.B. Reddy, Inhibition of aldose reductase by dietary antioxidant curcumin: Mechanism of inhibition, specificity and significance, FEBS Lett. 583 (2009) 3637–3642.

[104] R. Yan, X. Zu, J. Ma, Z. Liu, M. Adeyanju, D. Cao, Aldo-keto reductase family 1 B10 gene silencing results in growth inhibition of colorectal cancer cells: Implication for cancer intervention, Int. J. Cancer 121 (2007) 2301–2306.

[105] T. Matsunaga, S. Endo, M. Soda, H.T. Zhao, O. El-Kabbani, K. Tajima, et al., Potent and selective inhibition of the tumor marker AKR1B10 by bisdemethoxycurcumin: Probing the active site of the enzyme with molecular modeling and site-directed mutagenesis, Biochem. Biophys. Res. Commun. 389 (2009) 128–132.

[106] I.S. Kil, S.W. Shin, H.S. Yeo, Y.S. Lee, J.W. Park, Mitochondrial $NADP^+$-dependent isocitrate dehydrogenase protects cadmium-induced apoptosis, Mol. Pharmacol. 70 (2006) 1053–1061.

[107] Y.M. Lee, S.H. Park, D.I. Shin, J.Y. Hwang, B. Park, Y.J. Park, Oxidative modification of peroxiredoxin is associated with drug-induced apoptotic signaling in experimental models of Parkinson disease, J. Biol. Chem. 283 (2008) 9986–9998.

[108] A. Shehzad, J.W. Park, J. Lee, Y.S. Lee, Curcumin induces radiosensitivity of in vitro and in vivo cancer models by modulating pre-mRNA processing factor 4 (Prp4), Chem. Biol. Interact. 206 (2013) 394–402.

[109] S. Awasthi, U. Pandya, S.S. Singhal, J.T. Lin, V. Thiviyanathan, W.E. Seifert, Y.C. Awasthi, G.A. Ansari, Curcumin-glutathione interactions and the role of human glutathione S-transferase P1-1, Chem. Biol. Interact. 128 (2000) 19–38.

[110] A. Sun, Y.J. Lu, H. Hu, M. Shoji, D.C. Liotta, J.P. Snyder, Curcumin analog cytotoxicity against breast cancer cells: Exploitation of a redox-dependent mechanism, Bioorg. Med. Chem. Lett. 19 (2009) 6627–6631.

[111] R.A. Sharma, H.R. McLelland, K.A. Hill, C.R. Ireson, S.A. Euden, M.M. Manson, M. Pirmohamed, L.J. Marnett, A.J. Gescher, W.P. Steward, Pharmacodynamic and pharmacokinetic study of oral Curcuma extract in patients with colorectal cancer, Clin. Cancer Res. 7 (2001) 1894–1900.

[112] G. Garcea, D.P. Berry, D.J. Jones, R. Singh, A.R. Dennison, P.B. Farmer, R.A. Sharma, W.P. Steward, A.J. Gescher, Consumption of the putative chemopreventive agent curcumin by cancer patients: Assessment of curcumin levels in the colorectum and their pharmacodynamic consequences, Cancer Epidemiol. Biomarkers Prev. 14 (2005) 120–125.

[113] R.E. Carroll, R.V. Benya, D.K. Turgeon, S. Vareed, M. Neuman, L. Rodriguez, et al., Phase IIa clinical trial of curcumin for the prevention of colorectal neoplasia, Cancer Prev. Res. (Phila.) 4 (2011) 354–364.

[114] Z.Y. He, C.B. Shi, H. Wen, F.L. Li, B.L. Wang, J. Wang, Upregulation of p53 expression in patients with colorectal cancer by administration of curcumin, Cancer Invest. 29 (2011) 208–213.

[115] S. Shishodia, M.M. Chaturvedi, B.B. Aggarwal, Role of curcumin in cancer therapy, Curr. Probl. Cancer 31 (2007) 243–305.

[116] P. Anand, C. Sundaram, S. Jhurani, A.B. Kunnumakkara, B.B. Aggarwal, Curcumin and cancer: An "old-age" disease with an "age-old" solution, Cancer Lett. 267 (2008) 133–164.

[117] N. Dhillon, B.B. Aggarwal, R.A. Newman, S. Vareed, M. Neuman, L. Rodriguez, et al., Phase II trial of curcumin in patients with advanced pancreatic cancer, Clin. Cancer Res. 14 (2008) 4491–4499.

[118] J.L. Capodice, P. Gorroochurn, A.S. Cammack, G. Eric, J.M. McKiernan, M.C. Benson, B.A. Stone, A.E. Katz, Zyflamend in men with high-grade prostatic intraepithelial neoplasia: Results of a phase I clinical trial, J. Soc. Integr. Oncol. 7 (2009) 43–51.

[119] A.L. Cheng, C.H. Hsu, J.K. Lin, M.M. Hsu, Y.F. Ho, T.S. Shen, et al., Phase I clinical trial of curcumin, a chemopreventive agent, in patients with high-risk or premalignant lesions, Anticancer Res. 21 (2001) 2895–2900.

[120] T. Golombick, T.H. Diamond, V. Badmaev, A. Manoharan, R. Ramakrishna, The potential role of curcumin in patients with monoclonal gammopathy of undefined significance—Its effect on paraproteinemia and the urinary N-telopeptide of type I collagen bone turnover marker, Clin. Cancer Res. 15 (2009) 5917–5922.

[121] T. Golombick, T.H. Diamond, A. Manoharan, R. Ramakrishna, Monoclonal gammopathy of undetermined significance, smoldering multiple myeloma, and curcumin: A randomized, double-blind placebo-controlled cross-over 4 g study and an open-label 8 g extension study, Am. J. Hematol. 87 (2012) 455–460.

[122] S.G. Kim, M.S. Veena, S.K. Basak, E. Han, T. Tajima, D.W. Gjertson, et al., Curcumin treatment suppresses IKKbeta kinase activity of salivary cells of patients with head and neck cancer: A pilot study, Clin. Cancer Res. 17 (2011) 5953–5961.

[123] B. Rai, J. Kaur, R. Jacobs, J. Singh, Possible action mechanism for curcumin in pre cancerous lesions based on serum and salivary markers of oxidative stress, J. Oral Sci. 52 (2010) 251–256.
[124] N. Chainani-Wu, E. Madden, F. Lozada-Nur, S. Silverman, High-dose curcuminoids are efficacious in the reduction in symptoms and signs of oral lichen planus, J. Am. Acad. Dermatol. 66 (2012) 752–760.

CHAPTER NINE

Molecular Targets of Honokiol: A Promising Phytochemical for Effective Cancer Management

Courey Averett*, Sumit Arora*, Haseeb Zubair*, Seema Singh*, Arun Bhardwaj*, Ajay P. Singh*,†,1

*Department of Oncologic Sciences, Mitchell Cancer Institute, University of South Alabama, Mobile, Alabama, USA

†Department of Biochemistry and Molecular Biology, College of Medicine, University of South Alabama, Mobile, Alabama, USA

[1]Corresponding author: e-mail address: asingh@health.southalabama.edu

Contents

Abstract

Honokiol is a bioactive, biphenolic phytochemical, present in the aerial parts of plants classified under the genus *Magnolia*. It has been an important constituent of Asian traditional medicine and is used against many ailments. Honokiol possesses potent

The Enzymes, Volume 36
ISSN 1874-6047
http://dx.doi.org/10.1016/B978-0-12-802215-3.00009-4

antioxidative, anti-inflammatory, antiangiogenic, and anticancer activities by targeting a variety of signaling molecules. Consequently, there has been immense interest in exploring its utility as a novel chemopreventive and therapeutic agent against several malignancies. In this chapter, we review the structure–function relationship of honokiol and its derivative compounds, the impact of honokiol on various phenotypes associated with cancer progression and metastasis, and its prominent molecular targets and pharmacokinetics. Clearly, the available data generate significant interest in this novel phytochemical and emerging information continues to provide strong support for its potential applicability in cancer management.

1. INTRODUCTION

Cancer development is a multistage process involving a series of events occurring over an extended period of time. Its evolution is mediated through multiple genetic and epigenetic aberrations that lead to the changes in the signaling profile of the malignant cell as well as its microenvironment. A quick review of the literature provides considerable evidence to support that phytochemicals can inhibit cancer progression by targeting signaling molecules directly and/or by altering gene expression through epigenetic or transcriptomic modulation. Intake of plant-derived products has been associated with decreased cancer incidence, and 60% of all anticancer drugs being used today are either natural products or their modified forms [1,2]. Polyphenols are one of the important classes of phytochemicals demonstrated to possess anticancer properties. They are synthesized as secondary metabolites by plants as a defense mechanism against microorganisms, insects, and herbivores. Honokiol, a polyphenol isolated from the plants of the genus *Magnolia* (spp. *officinalis*, *obovata*, and *grandiflora*) of Magnoliaceae family has gained significant attention in recent years [3–5]. The genus *Magnolia* is distributed throughout the American continent as well as east and south-east Asia. Different parts of *Magnolia officinalis* have been used in traditional Chinese medicine and are often a component of Asian herbal teas locally known as *houpo*, *koboku*, or *saiboku-tu* [6]. A large number of studies have demonstrated diverse biological properties of honokiol such as antioxidative [7], antiarrhythmic [8], anxiolytic [9], antithrombocytic [10], anti-inflammatory [11], antiangiogenic [12–14], and antitumor effects [12,15,16]. Here, we summarize and discuss the literature on the anticancer properties, the important molecular targets, and pharmacological attributes of honokiol which support its promise for improving human health.

2. HONOKIOL: STRUCTURE–ACTIVITY RELATIONSHIP

Honokiol consists of *para*-allyl-phenol and *ortho*-allyl-phenol groups linked together with *ortho*- and *para*-C-C-coupling [3,16] (Fig. 9.1). Structure–activity relationship studies have demonstrated that the hydroxyl and allylic groups are responsible for the biological activities of honokiol. The 4′-phenol and 5-allyl groups are essential for honokiol's neurotrophic activity [3], and the 3′ allyl group has been suggested to play an important role in inducing its cytotoxic property [17]. Due to its relatively low solubility in oil and water, yet high-medicinal properties, a number of derivatives have been synthesized to enhance its activity. Novel derivatives of honokiol have been evaluated to possess neurotrophic [18], antimicrobial [19], and antitumor [14,20] activities. Three derivatives of honokiol, namely, 3′-formylhonokiol, 5-formylhonokiol, and 3′,5-diformylhonokiol have been demonstrated to possess potent antitumor properties (Fig. 9.1). Among them, 5-formylhonokiol possesses the strongest inhibitory activity against tumor cell lines K562 (human myelogenous leukemia), A549, and SPC-A1 (human lung adenocarcinoma) [6]. In another study on honokiol derivatives, 3′,5-diallyl-2,4′-dihydroxy-[1,1′-biphen-yl]-3,5′-dicarbaldehyde was found to significantly suppress angiogenesis *in vitro* and *in vivo*. This derivative had enhanced inhibitory effects on human umbilical vein endothelial cells (HUVECs), A549, HepG2 (hepatocellular carcinoma), and LL/2 (mouse Lewis lung carcinoma) cells as compared to honokiol, and was also shown to suppress newly grown segmental vessels from the dorsal aorta of zebrafish [14].

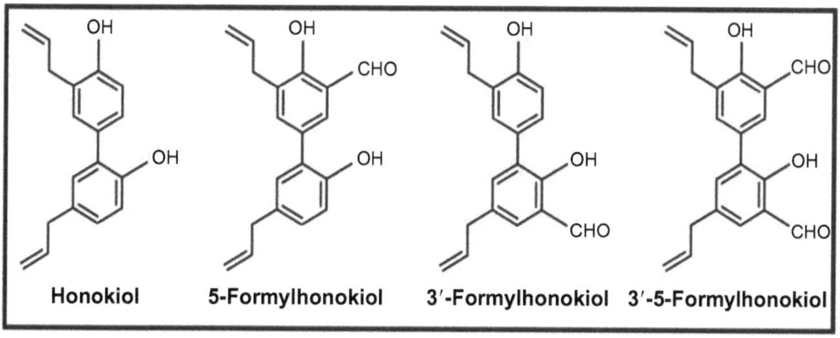

Figure 9.1 Chemical structure of honokiol and its anticancer derivatives, namely, 3′-formylhonokiol, 5-formylhonokiol, and 3′,5-diformylhonokiol. 3′-Formylhonokiol and 5-formylhonokiol are isomers.

3. ANTICANCER EFFECT OF HONOKIOL

The development of cancer is an extremely complex process that requires stimulation of growth, evasion of apoptosis, uncontrolled replication followed by angiogenesis, tissue invasion, and metastasis [21]. These processes represent prime targets for therapeutic interventions, and a potential anticancer agent should be able to interfere with one or more of these stages of cancer pathogenesis [22]. Honokiol has been shown to have growth inhibitory activity against various cancer cell lines including renal cell carcinoma [23], oral squamous cancer [24,25], leukemia [15,26], malignant melanoma [27], colon cancer cells [28,29], breast cancer [4,30,31], ovarian cancer [32,33], and pancreatic cancer [5]. In the following section, we summarize the effects of honokiol on the survival, development, and progression of cancer.

3.1. Cell-Cycle Arrest

Alterations in the regulation of the cell cycle have been found responsible for the continuous proliferation of cancer cells, and chemical compounds capable of impeding cell-cycle specific proteins can block the proliferation of such tumorigenic cells [34]. The cell cycle is regulated by concerted actions of cyclins, cyclin-dependent kinases (Cdks), and Cdk inhibitors. Cyclin D1 and its catalytic partner Cdk4 dominate in G1 phase, whereas cyclin E and Cdk2 complex regulates the cell-cycle progression from G1 to S [35]. It has been observed that treatment with honokiol resulted in G1-phase arrest of cell-cycle progression in cancer cells including breast cancer [4,36], leukemia [37], prostate cancer [38], skin cancer [39], and pancreatic cancer [5]. It downregulates the expression of cyclins (D1, D2, and E), cyclin-dependent kinases (Cdk2, Cdk4, and Cdk6) and enhances the expression of Cdk inhibitor proteins p21 and p27 [5,39]. Honokiol treatment also modulated the downregulation of retinoblastoma protein expression and led to the cleavage of poly (ADP-ribose) polymerase (PARP) and an increase in Bcl-xL to promote apoptosis in cancer cells [38].

3.2. Apoptosis Induction

Another major attribute for the survival of cancer is evasion of apoptosis. The effects of honokiol on apoptosis in different cancer types have been investigated. Honokiol induced formation of apoptotic bodies and generation of oligonucleosomal-sized DNA fragments, a hallmark feature of apoptosis in cancer cells [40]. The apoptosis-regulating Bcl-2 family comprises two functional groups, antiapoptotic Bcl-2 and Bcl-xL proteins and proapoptotic Bax, Bak, and Bad proteins, which provide the signal for cell survival or cell death through the activation of downstream effector caspase proteins. It has been shown that honokiol induced apoptosis through the induction of proapoptotic proteins Bax, Bak, and Bad, along with a decrease in the antiapoptotic proteins Bcl-xL and Mcl-1 [41,42]. Moreover, tumors from honokiol-treated mice exhibited a higher number of apoptotic bodies when compared to control tumors [41]. Activation of caspase-3, -8, -9, and the cleavage of PARP have also been demonstrated as an underlying mechanism of apoptosis induction by honokiol in B-cell chronic lymphocytic leukemia and in prostate cancer cells [26].

3.3. Antiangiogenic Effect

In order to facilitate the growing demand of nutrients of the developing tumor, cancer cells induce the formation of new blood vessels at the site of their growth, termed angiogenesis. Angiogenesis involves the induction of a large number of molecules, such as cytokines. These molecules are expressed by cancer cells, stromal cells, and tumor-associated macrophages in a concerted effort which leads to angiogenesis [43]. Honokiol is able to inhibit this process, and the mechanism of its antiangiogenic effect is mainly attributed to its ability to interfere with the phosphorylation of vascular endothelial growth factor receptor 2 (VEGFR2) on endothelial cells [12]. Using transformed endothelial cell line SVR, Bai *et al.* [12] have shown that honokiol exhibits potent inhibitory activity in proliferation assays. Furthermore, administration of tumor-bearing animals with honokiol inhibited angiogenesis and reduced tumor growth of ovarian cancer cells [13]. Honokiol also inhibited the vessel formation of mammalian embryonic stem cells on 3D collagen gel [44].

3.4. Inhibition of Migration and Invasion

Tumor cell migration and invasion are inter-related processes requiring the activities of proteolytic enzymes, including matrix metalloproteinases (MMPs such as MMP-2 and MMP-9) which degrade the extracellular matrix, basement membranes, and adhesion molecules (intercellular adhesion molecule, ICAM, and vascular cell adhesion molecule). Transition from an epithelial-to-mesenchymal (EMT) phenotype facilitates the breakdown of extracellular matrix followed by subsequent invasion of the surrounding tissues in order to enter the bloodstream and/or lymph nodes and travel to distant organ sites. Once cells have reached the distant organ sites, they undergo mesenchymal-to-epithelial transition and begin establishment of distal metastasis by the surviving cancer cells followed by outgrowth of secondary tumors [45]. Honokiol has been shown to inhibit the invasion of HT-1080 human fibrosarcoma cells and was also found to inhibit the expression of MMP-9, thereby inhibiting the metastatic potential of HT-1080 and U937 leukemia cells [12]. Another proposed mechanism for the inhibitory effects of honokiol on invasion and metastasis is through the liver kinase B1 (LKB1)/adenine monophosphate-activated protein kinase (AMPK) axis. Honokiol treatment increased the expression and cytoplasmic translocation of tumor-suppressor LKB1 in breast cancer cells, which led to the phosphorylation and functional activation of AMPK and resulted in the inhibition of invasion and metastasis [46]. Thus, the above studies indicate that honokiol can alter all the important steps of cancer progression and can be an effective therapeutic agent against cancer.

4. MOLECULAR TARGETS OF HONOKIOL

Cancer development and progression is a result of the dysregulation of multiple signaling pathways. Various studies have demonstrated that honokiol is able to target important molecules involved in these pathways to exert its anticancer effects. In this section, we will discuss the activity of honokiol against some of the most important molecular targets in signaling pathways that have been shown to influence cancer development and progression (Fig. 9.2).

4.1. Signal Transducers and Activators of Transcription

Signal transducers and activators of transcription (STATs) are a group of seven transcription factors that regulate cellular growth, cell cycle, survival, and angiogenesis [47]. In various malignancies, STATs are aberrantly activated due to functional loss of their negative regulators, or overexpression of upstream tyrosine kinases [48]. Rajendran *et al.* [47] demonstrated that honokiol inhibited the constitutive and inducible activation of STAT3 in hepatocellular carcinoma cells. Honokiol also inhibited STAT3 phosphorylation in OE33 and Jh-EsoAd1 esophageal adenocarcinoma cell lines, MDA-MB-231 breast cancer cells, and STAT3/5 phosphorylation in the MT-2T-cell leukemia cell line [36,49] in a dose- and time-dependent manner. A reduction in the STAT phosphorylation occurs within 30 min and

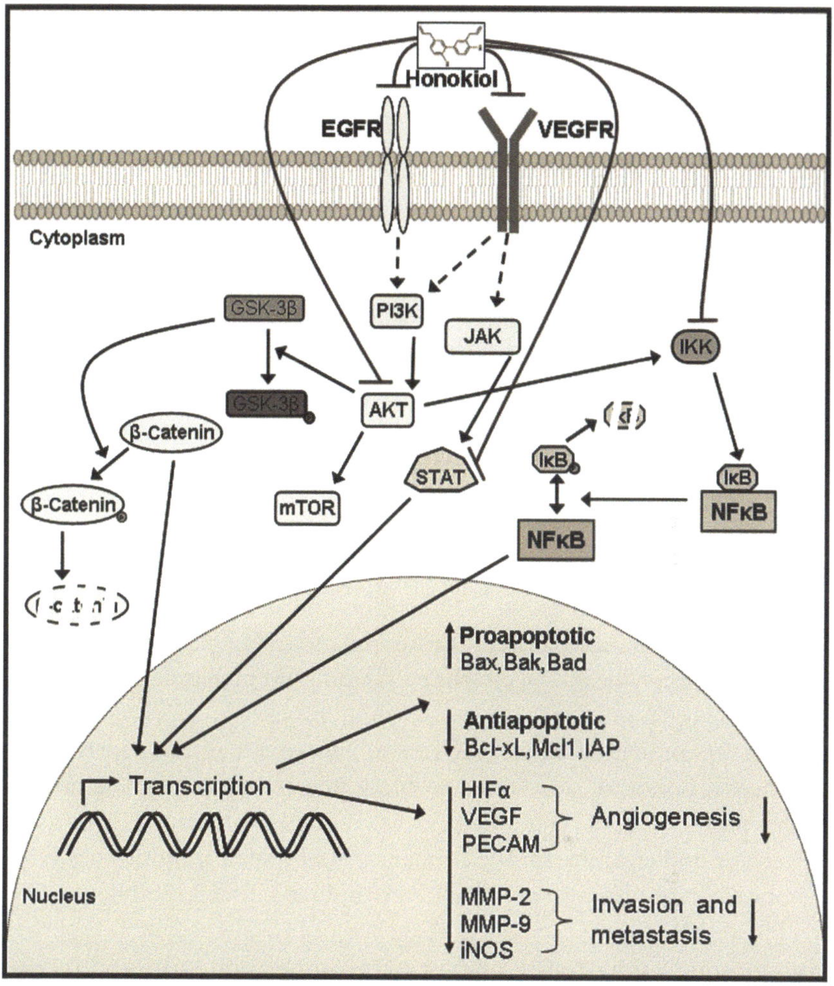

Figure 9.2 Important molecular targets of honokiol in tumor and/or endothelial cells and their involvement in cancer-relevant signaling pathways. Honokiol has been shown to influence (directly or indirectly) the levels and/or activity of several signaling molecules including signal transducers and activator of transcription (STATs), nuclear factor kappa B (NFκB), beta-catenin (β-catenin), mammalian target of rapamycin (mTOR), epidermal growth factor receptor (EGFR), and vascular endothelial growth factor receptor (VEGFR), which ultimately results in altered expression of genes involved in cell survival, proliferation, angiogenesis, invasion, and migration. (See the color plate.)

maximum inhibition is observed in 24 h [49,50]. Moreover, honokiol treatment decreased STAT3 overexpression and phosphorylation, leading to decreased levels of the mesenchymal transcription factor Zeb1 and increased E-cadherin levels in breast cancer cells [50]. Furthermore, SHP-1 (a tyrosine phosphatase inhibitor of STATs) protein expression was found to be augmented by honokiol *in vitro* as well as *in vivo*, which suggested that honokiol can induce endoplasmic reticulum stress to elicit SHP-1 activity in gastric cancer cells [50].

4.2. Nuclear Factor Kappa B

The nuclear factor kappa B (NFκB) family consists of five DNA-binding proteins (p65, p50, p52, cRel, and RelB) that differentially modulate gene transcription [51]. NFκB activation is involved in many cellular processes including proliferation, apoptosis, inflammation, angiogenesis, invasion, and migration [52]. Transcription factor NFκB is constitutively activated in multiple malignancies and has been pathologically implicated in various cancers [51]. Honokiol can alter the constitutive activation of NFκB and expression of NFκB-regulated gene products involved in apoptosis (IAP1, IAP2, Bcl-xL, Bcl-2, cFLIP, TRAF1, and survivin), proliferation (cyclin D1 and c-myc), inflammation (cyclooxygenase-2, COX-2), invasion (MMP-9 and ICAM-1), and angiogenesis (VEGF), thereby enhancing apoptosis and suppressing proliferation and invasion [53]. Inhibition of NFκB by honokiol has been shown in multiple cell types, including embryonic kidney cells, T-cell leukemia, multiple myeloma, lung adenocarcinoma, squamous cell carcinoma [53], pancreatic cancer [5], breast cancer [30], and colon cancer [29] cells. Moreover, in lung adenocarcinoma cells, honokiol was found to block the induced NFκB activation in the presence of TNFα, okadaic acid, H_2O_2, phorbol 12-myristate 13-acetate, and cigarette smoke condensate [53]. Such an inhibitory effect of honokiol was also observed in TNFα-induced T-cell lymphoma cells. IκB kinases (IKKs) are the master regulator of NFκB-mediated responses. IKKs are rapidly activated in response to cell stimulation and phosphorylate two critical serines in the N-terminal regulatory domain of the IκBα. Phosphorylated IκBα undergoes polyubiquitination and rapid degradation leading to the release of NFκB dimers that translocate to the nucleus to activate gene transcription [54]. Honokiol treatment also resulted in a dose-dependent decrease in activation of IKKs (i.e., IKKα and IKKβ) [53], leading to increased IκBα levels and increased cytoplasmic sequestration of NFκB.

4.3. Beta-Catenin

Beta-catenin (β-catenin) is a multifunctional, 90 kD protein that contributes to cell development under normal physiological conditions [55]. β-Catenin is a crucial transcriptional factor in Wingless–Int (Wnt) signaling, and plays important role in stem cell renewal and organ regeneration [56]. Aberrant expression of β-catenin induces malignant transformation of normal cells, and its abnormal activity has been reported in many cancer types [57]. Normally, cytoplasmic β-catenin levels are kept low through continuous, phosphorylation-induced proteosome-mediated degradation by a protein inhibitory complex comprising of adenomatous polyposis coli/glycogen synthase kinase-3β (GSK-3β)/axin [58]. Upon activation of Wnt signaling, the GSK-3β is phosphorylated followed by the dissociation of the inhibitory complex. Thus, degradation of β-catenin is inhibited and levels of β-catenin increase in the cytoplasm, which then translocation to nucleus. β-Catenin interacts with the transcription factor, T-cell factor/lymphoid enhancer-binding factor, and regulates the transcription of various target genes associated with cell survival, proliferation, metastasis, and EMT [58]. Honokiol has been examined to have inhibitory effects on β-catenin. In non-small cell lung cancer cells, honokiol reduced the nuclear accumulation of β-catenin as well as increased the phosphorylation of β-catenin [31] due to the increased stability of β-catenin regulatory complex (CK1α and GSK-3β). In a study by Yao *et al.* [59], honokiol suppressed growth and clonogenicity of oral squamous cell carcinoma cells and the side population cells through the reduction of β-catenin and an increase in GSK-3α/β protein levels. Moreover, downstream targets of β-catenin *viz* c-Myc and cyclin D1 were also found to be suppressed in these honokiol-treated cells. Honokiol has also been shown to reduce the invasiveness of glioblastoma cells through inhibition of β-catenin expression [60].

4.4. Phosphoinositide 3-Kinase/Akt/Mammalian Target of Rapamycin

The phosphoinositide 3-kinase (PI3K)/Akt/mammalian target of rapamycin (mTOR) signaling pathway is crucial to several functions involving cell growth, cell-cycle progression, and therapy resistance in cancer [30]. These signaling pathways are so interconnected that it can be considered as a single, unique pathway. The PI3K/Akt pathway is responsible for the abnormal activation of mTOR in cancer cells. The efficacy of honokiol on the PI3K/Akt/mTOR pathway has been examined in a variety of breast cancer cell lines, and it has been demonstrated that honokiol decreased Akt activation through the decreased phosphorylation of Akt [30]. Furthermore, coadministration of honokiol with oxaliplatin synergistically abolished

Akt signaling in colon cancer cells [29]. It was also observed that the cotreatment of honokiol with rapamycin (a pharmacological inhibitor of mTOR) resulted in a more dramatic induction of apoptosis on the Her-2 negative cells [30]. Honokiol attenuated signaling of the PI3K/Akt/mTOR pathway in PTEN-deficient breast, prostate, and glioma cell lines [61], and Ishikawa *et al.* [37] reported similar results in T-cell leukemia MT-2 cell lines after honokiol treatment. The currently known pharmacological inhibitors targeting PI3K/Akt/mTOR signaling pathway have been found to significantly impede stimulated T-cell proinflammatory function, hampering T-cell immunotherapy. However, honokiol treatment did not alter the growth of stimulated T-cells or proinflammatory function at PI3K/mTOR inhibitory dose in cancer cells [61]. In addition to reducing phosphorylation of Akt, honokiol was found to inhibit phosphorylation of both the upstream mediator c-Src and downstream effector mTOR in breast cancer cells [36].

4.5. Epidermal Growth Factor Receptor

The epidermal growth factor receptor (EGFR) is a member of the ErbB family of receptor tyrosine kinases and is upregulated in many cancer cell types, including breast cancer and head and neck squamous cell carcinoma (HNSCC) [36,62]. When activated by its ligand, EGFR activates signaling cascades involving STATs, MAPK, and PI3K/Akt. Thus, EGFR is a critical mediator of cellular proliferation and angiogenesis. In a study on breast cancer cell lines, honokiol induced the suppression of EGFR protein expression. Furthermore, heat shock protein 90, a protein chaperone responsible to stabilize EGFR, was also downregulated by honokiol [36]. Leeman-Neill and coworkers [62] demonstrated a remarkable decrease in EGFR protein, as well as phosphorylated EGFR, after the treatment of HNSCC Cal-33 cells with honokiol, thus suggesting that honokiol affects the total protein level and its activation. Similarly, Wang *et al.* observed that honokiol attenuated paclitaxel chemoresistance through both inhibition of EGFR phosphorylation and decreased expression of EGFR protein in multidrug resistant, lineage-derived cervical cancer cell lines KB-8-5, KB-C-1, and KB-V1 [63]. Furthermore, the administration of honokiol in an HNSCC xenograft model, alone or in combination with cetuximab (a tyrosine kinase inhibitor), led to a remarkable reduction in EGFR levels in the honokiol-treated group as compared to cetuximab alone. They also found that honokiol significantly enhanced the inhibitory effect of cetuximab, which resulted in reduced tumor volume in the combination group [62].

4.6. Vascular Endothelial Growth Factor and Its Receptor

Tumor angiogenesis is an essential process for the continuous growth of a tumor and subsequent metastases [21]. Vascular endothelial growth factor (VEGF) is a signal protein that stimulates blood vessel formation; its receptor, vascular endothelial growth factor receptor (VEGFR), is involved in both the formation of the circulatory system (during embryogenesis) as well as the growth of blood vessels from preexisting vasculature [64]. Recent reports show that VEGF is also an important player in angiogenesis and the development of chemoresistance [65]. Thus, the possibility of targeting of VEGF/VEGFR can have broad therapeutic implications against cancer. The effect of honokiol on VEGF production and VEGFR isoforms has been assessed in multiple cancer cell lines as well as *in vivo* models. It has been found that honokiol abrogates the induction of VEGF in response to TNF treatment [53] and interferes with the phosphorylation of VEGFR-2 [12]. In vascular endothelial cells, the level of VEGF was observed to be lower in honokiol-treated cells as compared to control cells [66]. Honokiol downregulated VEGF-D (a VEGF family member) production in Lewis lung carcinoma cells, and it was also found to reduce the expression of VEGFR-2 and VEGFR-3 in endothelial cells (HUVECs and HLECs) [66]. In colon cancer cells treated with honokiol, there was a significant reduction in VEGF expression in cell culture supernatants as well as in cellular lysates [29]. Significantly low VEGF expression was observed after treatment with honokiol in an *in vivo* ovarian cancer model [13]. In a study investigating the effect of honokiol on retinal epithelial cells, it was found that under both hypoxic and normoxic conditions, honokiol treatment resulted in a reduction in the mRNA levels of VEGF [67]. VEGF-mediated mitogenic and chemotactic effects on cultured HUVEC cells were predominately mediated through VEGF2 tyrosine kinase, KDR. Altogether, honokiol was found to markedly inhibit VEGF-induced response in a dose-dependent manner [12].

4.7. Hypoxia-Inducible Factors

Hypoxia-inducible factors (HIFs) are master regulators of cancer cells' response in an oxygen-deficient environment (hypoxia) to sustain their survival and growth. HIFs activate the transcription of genes involved in cancer pathobiology such as cell proliferation, angiogenesis, metabolism, and invasion. Microenvironments surrounding cancerous tumors are hypoxic, and proliferation of cancer cells in such areas requires HIF-1 activation. The inhibitory effect of honokiol on genes controlled by hypoxia-responsive elements has been observed as a result of the reduction in HIF-1α and HIF-2α protein levels *in vitro* and *in vivo* under hypoxic and normoxic conditions [68]. In a study conducted by Vavilala *et al.* [67] using multiple cell lines including human retinal pigment epithelial cells, breast adenocarcinoma (MCF-7), colon adenocarcinoma (HT-29), and human embryonic kidney cells, they demonstrated that honokiol reduced the mRNA levels of HIF-1α, HIF-2α, and HIF-1β under hypoxic conditions. In addition, a marked decrease in histone lysine demethylases, specifically JMJD1A, JMJD2B, JMJD2C, and JAIRD1B, was observed in presence of honokiol [67]. Furthermore, in a recent study it was demonstrated that honokiol effectively blocked the expression of stem cell markers in hemangioblastoma cells that are under the influence of HIF pathway [69].

4.8. Cyclooxygenases

Cyclooxygenase (COX) enzymes, COX-1 and COX-2, are responsible for the catalysis of prostaglandin synthesis from its precursor arachidonic acid. COX-2 is normally undetectable in healthy tissue, yet increased amounts of COX-2 can be detected in premalignant as well as in malignant tissue [70]. In gastric cancer, Liu *et al.* [71] found that COX-2 and peroxisome proliferator-activated receptor gamma are overexpressed in gastric cancer cells and xenograft mice model, which can be suppressed by treatment with honokiol to inhibit tumor growth and cancer cell viability. This inhibition of COX-2 did not affect COX-1 protein expression, which suggests that there is a higher specificity by honokiol than NSAIDs, which are known to be less selective toward COX subtypes. Further, they determined that honokiol treatment resulted in upregulation of anticarcinogenic lipoxygenases in colorectal cancer [71]. In a study on lung adenocarcinoma cells, it was demonstrated that COX-2 expression induced by TNF can be effectively suppressed by honokiol [53]. In colon cancer cell line HT-29, researchers found that honokiol-inhibited COX-2 protein expression which resulted in reduced PGE2 in cell culture supernatants [29]. In breast cancer cell lines MCF-7 and 4T1, Singh and Katiyar [31] demonstrated that breast cancer cell migration associated with COX-2 is inhibited with honokiol treatment. They further observed that there was a reduction of nitric oxide in a dose-dependent manner after treatment with honokiol. Furthermore, a reduced migratory and invasive capacity of breast cancer cells was also observed following inducible nitric oxide synthase inhibition.

5. PHARMACOKINETICS OF HONOKIOL

Pharmacokinetics involves the movement of drug into, through, and out of the body. It includes the time course of absorption, distribution, metabolism, and excretion. There are only a few studies on the pharmacokinetics of honokiol. Collectively, these studies support that honokiol undergoes a rapid distribution and slow elimination after intravenous (i.v.) administration. In a pharmacokinetic study using i.v. administration of a single dose of free honokiol in Sprague-Dawley rats, a rapid distribution phase followed by a slower elimination phase (elimination $t_{1/2}=49.22$ and 56.2 min for 5 or 10 mg of honokiol, respectively) was observed [72]. Recently, pharmacokinetic parameters of honokiol in beagle dogs' (male and female) plasma were determined for the first time after i.v. administration [23]. Elimination half-life ($t_{1/2}$ in hours) was found to be 20.13 (female), 9.27 (female), 7.06 (male), 4.70 (male), and 1.89 (male) after administration of doses (mg/kg) 8.8, 19.8, 3.9, 44.4, and 66.7, respectively. The $t_{1/2}$ decreases with an increase in the dose and length of infusion. Chen *et al.* [73] investigated the pharmacokinetics of honokiol after intraperitoneal injection of 250 mg/kg in BALB/c mice and observed the maximum plasma concentration of honokiol at 27.179 ± 6.252 min, with the $t_{1/2}$ of 312.08 ± 51.66 min. In another study on Wistar rats, the $t_{1/2}$ was found to be 526.63 min after a single oral dose of honokiol decoction (5 g/kg body weight), which suggested that rhubarb and immature orange fruit in decoction influence the pharmacokinetics of honokiol [74]. The maximal plasma concentrations of honokiol following rectal administration of Houpo extract at a dose of 245 mg/kg was approximately six times that of an identical oral dose, indicating that rectal dosing led to inhibition of first-pass metabolism [75].

Bohmdorfer *et al.* [76] investigated the metabolism of honokiol in rat and human livers using liver microsomes, liver cytosol, and isolated perfused rat liver. Their study suggests that sulfation and glucuronidation are the main metabolic pathways contributing to the clearance of honokiol *in vivo*. Wang *et al.* [77] developed polyethylene glycol-coated (PEGylated) liposomal honokiol with the purpose of improving the aqueous solubility and pharmacokinetics in comparison with free honokiol. The pharmacokinetic analysis of PEGylated liposomal honokiol in Balb/c mice showed a twofold increase in elimination $t_{1/2}$ value as compared to that of free honokiol. Moreover, $AUC_{0\rightarrow\infty}$ (mean concentration of drug in plasma) of PEGylated liposomal honokiol was about 1.85-fold higher than free honokiol. The protein-binding ability of honokiol in plasma was reported to be between 60% and 65%. In another study, plasma honokiol concentrations were observed to be above 30 and 10 μg/mL for 24 and 48 h, respectively, in liposomal honokiol-treated mice, whereas it fell quickly (<5 μg/mL) by 12 h in free honokiol-treated mice bearing A549 xenograft tumors suggesting that liposomal honokiol prolonged blood circulation times in tumor-bearing mice in comparison with free honokiol [78]. A recent study on honokiol distribution in plasma, brain, and other tissues showed the percentage of honokiol ordered as follows: lungs > plasma > liver > brain > kidney > heart > spleen in animal models (rat 9 L intracerebral gliosarcoma model in Fisher 344 rats and human U251 xenograft glioma model in nude mice) [77]. The pharmacokinetics profile of free honokiol can be concluded as follows: poor gastrointestinal absorption, approximately 60–65% is plasma protein bound, metabolization occurs in the liver (sulfation and glucuronidation), approximately 50% is secreted in bile, and elimination $t_{1/2}$ of 49.05–56.24 min.

6. CONCLUSION AND FUTURE OUTLOOK

There is clearly ample evidence to support the antitumor activity of honokiol. It has been shown to alter the growth, progression, metastasis, and angiogenesis and is bioactive against several malignancies. We now have the information available on its mechanisms of action and several important targets of honokiol have been characterized. More importantly, honokiol is nontoxic after systemic or oral administration and exhibits a desirable spectrum of pharmacokinetics, which further makes it an attractive natural agent for cancer prevention and therapy. Moreover, additional data from its derivative compounds suggest that there is still scope for improving its biological and pharmacological properties. We may also benefit from the advances in the area of nanotechnology to develop an oral formulation, a more desirable format for chemopreventive use than invasive delivery. Besides, it can also help in targeted delivery of honokiol to primary and metastatic sites to promote its antitumor efficacy and limiting any unforeseen side effects. It may be about time for basic, translational, and clinical researchers to join together to take this promising phytochemical to the next level by conducting clinical trials and addressing any short-comings on its way to human applications.

ACKNOWLEDGMENTS

We would like to acknowledge the funding support from NIH/NCI [CA137513, CA167137, CA175772, and CA185490 (to A. P. S.) and CA169829 and CA186233 (to S. S.)] and USAMCI.

REFERENCES

[1] D.J. Newman, G.M. Cragg, K.M. Snader, Natural products as sources of new drugs over the period 1981-2002, J. Nat. Prod. 66 (2003) 1022–1037.
[2] H. Vainio, E. Weiderpass, Fruit and vegetables in cancer prevention, Nutr. Cancer 54 (2006) 111–142.
[3] T. Esumi, G. Makado, H. Zhai, Y. Shimizu, Y. Mitsumoto, Y. Fukuyama, Efficient synthesis and structure-activity relationship of honokiol, a neurotrophic biphenyl-type neolignan, Bioorg. Med. Chem. Lett. 14 (2004) 2621–2625.
[4] I. Wolf, J. O'Kelly, N. Wakimoto, A. Nguyen, F. Amblard, B.Y. Karlan, J.L. Arbiser, H.P. Koeffler, Honokiol, a natural biphenyl, inhibits in vitro and in vivo growth of breast cancer through induction of apoptosis and cell cycle arrest, Int. J. Oncol. 30 (2007) 1529–1537.
[5] S. Arora, A. Bhardwaj, S.K. Srivastava, S. Singh, S. McClellan, B. Wang, A.P. Singh, Honokiol arrests cell cycle, induces apoptosis, and potentiates the cytotoxic effect of gemcitabine in human pancreatic cancer cells, PLoS One 6 (2011) e21573.
[6] Y. Luo, Y. Xu, L. Chen, H. Luo, C. Peng, J. Fu, H. Chen, A. Peng, H. Ye, D. Xie, A. Fu, J. Shi, S. Yang, Y. Wei, Preparative purification of anti-tumor derivatives of honokiol by high-speed counter-current chromatography, J. Chromatogr. A 1178 (2008) 160–165.
[7] C. Zhao, Z.Q. Liu, Comparison of antioxidant abilities of magnolol and honokiol to scavenge radicals and to protect DNA, Biochimie 93 (2011) 1755–1760.
[8] S.K. Tsai, C.H. Huang, S.S. Huang, L.M. Hung, C.Y. Hong, Antiarrhythmic effect of magnolol and honokiol during acute phase of coronary occlusion in anesthetized rats: Influence of L-NAME and aspirin, Pharmacology 59 (1999) 227–233.
[9] H. Kuribara, W.B. Stavinoha, Y. Maruyama, Honokiol, a putative anxiolytic agent extracted from magnolia bark, has no diazepam-like side-effects in mice, J. Pharm. Pharmacol. 51 (1999) 97–103.
[10] H. Hu, X.X. Zhang, Y.Y. Wang, S.Z. Chen, Honokiol inhibits arterial thrombosis through endothelial cell protection and stimulation of prostacyclin, Acta Pharmacol. Sin. 26 (2005) 1063–1068.
[11] K.T. Liou, S.M. Lin, S.S. Huang, C.L. Chih, S.K. Tsai, Honokiol ameliorates cerebral infarction from ischemia-reperfusion injury in rats, Planta Med. 69 (2003) 130–134.
[12] X. Bai, F. Cerimele, M. Ushio-Fukai, M. Waqas, P.M. Campbell, B. Govindarajan, C.J. Der, T. Battle, D.A. Frank, K. Ye, E. Murad, W. Dubiel, G. Soff, J.L. Arbiser, Honokiol, a small molecular weight natural product, inhibits angiogenesis in vitro and tumor growth in vivo, J. Biol. Chem. 278 (2003) 35501–35507.
[13] Z. Li, Y. Liu, X. Zhao, X. Pan, R. Yin, C. Huang, L. Chen, Y. Wei, Honokiol, a natural therapeutic candidate, induces apoptosis and inhibits angiogenesis of ovarian tumor cells, Eur. J. Obstet. Gynecol. Reprod. Biol. 140 (2008) 95–102.

[14] L. Ma, J. Chen, X. Wang, X. Liang, Y. Luo, W. Zhu, T. Wang, M. Peng, S. Li, S. Jie, A. Peng, Y. Wei, L. Chen, Structural modification of honokiol, a biphenyl occurring in Magnolia officinalis: The evaluation of honokiol analogues as inhibitors of angiogenesis and for their cytotoxicity and structure-activity relationship, J. Med. Chem. 54 (2011) 6469–6481.

[15] H. Hibasami, Y. Achiwa, H. Katsuzaki, K. Imai, K. Yoshioka, K. Nakanishi, Y. Ishii, M. Hasegawa, T. Komiya, Honokiol induces apoptosis in human lymphoid leukemia Molt 4B cells, Int. J. Mol. Med. 2 (1998) 671–673.

[16] S. Arora, S. Singh, G.A. Piazza, C.M. Contreras, J. Panyam, A.P. Singh, Honokiol: A novel natural agent for cancer prevention and therapy, Curr. Mol. Med. 12 (2012) 1244–1252.

[17] Z.L. Kong, S.C. Tzeng, Y.C. Liu, Cytotoxic neolignans: An SAR study, Bioorg. Med. Chem. Lett. 15 (2005) 163–166.

[18] K.V. Praveen, R.R. Gajendra, D.D. Vo, S. Chakravarty, S. Chandrasekhar, R. Gree, Synthesis and neurite growth evaluation of new analogues of honokiol, a neolignan with potent neurotrophic activity, Bioorg. Med. Chem. Lett. 22 (2012) 1439–1444.

[19] Y.S. Kim, J.Y. Lee, J. Park, W. Hwang, J. Lee, D. Park, Synthesis and microbiological evaluation of honokiol derivatives as new antimicrobial agents, Arch. Pharm. Res. 33 (2010) 61–65.

[20] Y. Luo, Y. Xu, L. Chen, J. Hu, C. Peng, D. Xie, J. Shi, W. Huang, G. Xu, M. Peng, J. Han, R. Li, S. Yang, Y. Wei, Semi-synthesis and anti-proliferative activity evaluation of novel analogues of Honokiol, Bioorg. Med. Chem. Lett. 19 (2009) 4702–4705.

[21] D. Hanahan, R.A. Weinberg, Hallmarks of cancer: The next generation, Cell 144 (2011) 646–674.

[22] G.L. Russo, Ins and outs of dietary phytochemicals in cancer chemoprevention, Biochem. Pharmacol. 74 (2007) 533–544.

[23] W. Li, Q. Wang, Q. Su, D. Ma, C. An, L. Ma, H. Liang, Honokiol suppresses renal cancer cells' metastasis via dual-blocking epithelial-mesenchymal transition and cancer stem cell properties through modulating miR-141/ZEB2 signaling, Mol. Cells 37 (2014) 383–388.

[24] X.R. Chen, R. Lu, H.X. Dan, G. Liao, M. Zhou, X.Y. Li, N. Ji, Honokiol: A promising small molecular weight natural agent for the growth inhibition of oral squamous cell carcinoma cells, Int. J. Oral Sci. 3 (2011) 34–42.

[25] D.W. Kim, S.M. Ko, Y.J. Jeon, Y.W. Noh, N.J. Choi, S.D. Cho, H.S. Moon, Y.S. Cho, J.C. Shin, S.M. Park, K.S. Seo, J.Y. Choi, J.I. Chae, J.H. Shim, Anti-proliferative effect of honokiol in oral squamous cancer through the regulation of specificity protein 1, Int. J. Oncol. 43 (2013) 1103–1110.

[26] T.E. Battle, J. Arbiser, D.A. Frank, The natural product honokiol induces caspase-dependent apoptosis in B-cell chronic lymphocytic leukemia (B-CLL) cells, Blood 106 (2005) 690–697.

[27] G. Kaushik, S. Ramalingam, D. Subramaniam, P. Rangarajan, P. Protti, P. Rammamoorthy, S. Anant, J.M. Mammen, Honokiol induces cytotoxic and cytostatic effects in malignant melanoma cancer cells, Am. J. Surg. 204 (2012) 868–873.

[28] S. Ponnurangam, J.M. Mammen, S. Ramalingam, Z. He, Y. Zhang, S. Umar, D. Subramaniam, S. Anant, Honokiol in combination with radiation targets notch signaling to inhibit colon cancer stem cells, Mol. Cancer Ther. 11 (2012) 963–972.

[29] H. Hua, W. Chen, L. Shen, Q. Sheng, L. Teng, Honokiol augments the anti-cancer effects of oxaliplatin in colon cancer cells, Acta Biochim. Biophys. Sin. (Shanghai) 45 (2013) 773–779.

[30] H. Liu, C. Zang, A. Emde, M.D. Planas-Silva, M. Rosche, A. Kuhnl, C.O. Schulz, E. Elstner, K. Possinger, J. Eucker, Anti-tumor effect of honokiol alone and in

combination with other anti-cancer agents in breast cancer, Eur. J. Pharmacol. 591 (2008) 43–51.
[31] T. Singh, S.K. Katiyar, Honokiol, a phytochemical from Magnolia spp., inhibits breast cancer cell migration by targeting nitric oxide and cyclooxygenase-2, Int. J. Oncol. 38 (2011) 769–776.
[32] L. Shayesteh, Y. Lu, W.L. Kuo, R. Baldocchi, T. Godfrey, C. Collins, D. Pinkel, B. Powell, G.B. Mills, J.W. Gray, PIK3CA is implicated as an oncogene in ovarian cancer, Nat. Genet. 21 (1999) 99–102.
[33] H. Luo, Q. Zhong, L.J. Chen, X.R. Qi, A.F. Fu, H.S. Yang, F. Yang, H.G. Lin, Y.Q. Wei, X. Zhao, Liposomal honokiol, a promising agent for treatment of cisplatin-resistant human ovarian cancer, J. Cancer Res. Clin. Oncol. 134 (2008) 937–945.
[34] T. Waldman, Y. Zhang, L. Dillehay, J. Yu, K. Kinzler, B. Vogelstein, J. Williams, Cell-cycle arrest versus cell death in cancer therapy, Nat. Med. 3 (1997) 1034–1036.
[35] L.H. Hartwell, M.B. Kastan, Cell cycle control and cancer, Science 266 (1994) 1821–1828.
[36] E.J. Park, H.Y. Min, H.J. Chung, J.Y. Hong, Y.J. Kang, T.M. Hung, U.J. Youn, Y.S. Kim, K. Bae, S.S. Kang, S.K. Lee, Down-regulation of c-Src/EGFR-mediated signaling activation is involved in the honokiol-induced cell cycle arrest and apoptosis in MDA-MB-231 human breast cancer cells, Cancer Lett. 277 (2009) 133–140.
[37] C. Ishikawa, J.L. Arbiser, N. Mori, Honokiol induces cell cycle arrest and apoptosis via inhibition of survival signals in adult T-cell leukemia, Biochim. Biophys. Acta 1820 (2012) 879–887.
[38] E.R. Hahm, S.V. Singh, Honokiol causes G0-G1 phase cell cycle arrest in human prostate cancer cells in association with suppression of retinoblastoma protein level/phosphorylation and inhibition of E2F1 transcriptional activity, Mol. Cancer Ther. 6 (2007) 2686–2695.
[39] C. Chilampalli, R. Guillermo, R.S. Kaushik, A. Young, G. Chandrasekher, H. Fahmy, C. Dwivedi, Honokiol, a chemopreventive agent against skin cancer, induces cell cycle arrest and apoptosis in human epidermoid A431 cells, Exp. Biol. Med. (Maywood) 236 (2011) 1351–1359.
[40] G.L. Yang, W.L. Hou, A.F. Fu, Z.H. Zhong, X.H. Yin, L.J. Chen, Effects of honokiol on proliferation and apoptosis of human cervical carcinoma cell line Hela in vitro, Sichuan Da Xue Xue Bao Yi Xue Ban 39 (2008) 558–562.
[41] E.R. Hahm, J.A. Arlotti, S.W. Marynowski, S.V. Singh, Honokiol, a constituent of oriental medicinal herb Magnolia officinalis, inhibits growth of PC-3 xenografts in vivo in association with apoptosis induction, Clin. Cancer Res. 14 (2008) 1248–1257.
[42] T. Wang, F. Chen, Z. Chen, Y.F. Wu, X.L. Xu, S. Zheng, X. Hu, Honokiol induces apoptosis through p53-independent pathway in human colorectal cell line RKO, World J. Gastroenterol. 10 (2004) 2205–2208.
[43] G. Bergers, L.E. Benjamin, Tumorigenesis and the angiogenic switch, Nat. Rev. Cancer 3 (2003) 401–410.
[44] G.D. Kim, S.Y. Bae, H.J. Park, K. Bae, S.K. Lee, Honokiol inhibits vascular vessel formation of mouse embryonic stem cell-derived endothelial cells via the suppression of PECAM and MAPK/mTOR signaling pathway, Cell. Physiol. Biochem. 30 (2012) 758–770.
[45] C.A. Klein, Cancer. The metastasis cascade, Science 321 (2008) 1785–1787.
[46] A. Nagalingam, J.L. Arbiser, M.Y. Bonner, N.K. Saxena, D. Sharma, Honokiol activates AMP-activated protein kinase in breast cancer cells via an LKB1-dependent pathway and inhibits breast carcinogenesis, Breast Cancer Res. 14 (2012) R35.
[47] P. Rajendran, F. Li, M.K. Shanmugam, S. Vali, T. Abbasi, S. Kapoor, K.S. Ahn, A.P. Kumar, G. Sethi, Honokiol inhibits signal transducer and activator of

transcription-3 signaling, proliferation, and survival of hepatocellular carcinoma cells via the protein tyrosine phosphatase SHP-1, J. Cell. Physiol. 227 (2012) 2184–2195.
[48] T. Bowman, R. Garcia, J. Turkson, R. Jove, STATs in oncogenesis, Oncogene 19 (2000) 2474–2488.
[49] C. Yu, Q. Zhang, H.Y. Zhang, X. Zhang, X. Huo, E. Cheng, D.H. Wang, J.L. Arbiser, S.J. Spechler, R.F. Souza, Targeting the intrinsic inflammatory pathway: Honokiol exerts proapoptotic effects through STAT3 inhibition in transformed Barrett's cells, Am. J. Physiol. Gastrointest. Liver Physiol. 303 (2012) G561–G569.
[50] D.B. Avtanski, A. Nagalingam, M.Y. Bonner, J.L. Arbiser, N.K. Saxena, D. Sharma, Honokiol inhibits epithelial-mesenchymal transition in breast cancer cells by targeting signal transducer and activator of transcription 3/Zeb1/E-cadherin axis, Mol. Oncol. 8 (2014) 565–580.
[51] H.J. Dyson, E.A. Komives, Role of disorder in IkappaB-NFkappaB interaction, IUBMB Life 64 (2012) 499–505.
[52] H.M. Shen, V. Tergaonkar, NFkappaB signaling in carcinogenesis and as a potential molecular target for cancer therapy, Apoptosis 14 (2009) 348–363.
[53] K.S. Ahn, G. Sethi, S. Shishodia, B. Sung, J.L. Arbiser, B.B. Aggarwal, Honokiol potentiates apoptosis, suppresses osteoclastogenesis, and inhibits invasion through modulation of nuclear factor-kappaB activation pathway, Mol. Cancer Res. 4 (2006) 621–633.
[54] M. Karin, M. Delhase, The I kappa B kinase (IKK) and NF-kappa B: Key elements of proinflammatory signalling, Semin. Immunol. 12 (2000) 85–98.
[55] W.H. Lien, E. Fuchs, Wnt some lose some: Transcriptional governance of stem cells by Wnt/beta-catenin signaling, Genes Dev. 28 (2014) 1517–1532.
[56] E. Trompouki, T.V. Bowman, L.N. Lawton, Z.P. Fan, D.C. Wu, A. DiBiase, C.S. Martin, J.N. Cech, A.K. Sessa, J.L. Leblanc, P. Li, E.M. Durand, C. Mosimann, G.C. Heffner, G.Q. Daley, R.F. Paulson, R.A. Young, L.I. Zon, Lineage regulators direct BMP and Wnt pathways to cell-specific programs during differentiation and regeneration, Cell 147 (2011) 577–589.
[57] P.J. Morin, Beta-catenin signaling and cancer, Bioessays 21 (1999) 1021–1030.
[58] R. Dajani, E. Fraser, S.M. Roe, M. Yeo, V.M. Good, V. Thompson, T.C. Dale, L.H. Pearl, Structural basis for recruitment of glycogen synthase kinase 3beta to the axin-APC scaffold complex, EMBO J. 22 (2003) 494–501.
[59] C.J. Yao, G.M. Lai, C.T. Yeh, M.T. Lai, P.H. Shih, W.J. Chao, J. Whang-Peng, S.E. Chuang, T.Y. Lai, Honokiol eliminates human oral cancer stem-like cells accompanied with suppression of Wnt/beta-catenin signaling and apoptosis induction, Evid. Based Complement. Alternat. Med. 2013 (2013) 146136, http://dx.doi.org/10.1155/2013/146136.
[60] J.J. Jeong, J.H. Lee, K.C. Chang, H.J. Kim, Honokiol exerts an anticancer effect in T98G human glioblastoma cells through the induction of apoptosis and the regulation of adhesion molecules, Int. J. Oncol. 41 (2012) 1358–1364.
[61] C. Crane, A. Panner, R.O. Pieper, J. Arbiser, A.T. Parsa, Honokiol-mediated inhibition of PI3K/mTOR pathway: A potential strategy to overcome immunoresistance in glioma, breast, and prostate carcinoma without impacting T cell function, J. Immunother. 32 (2009) 585–592.
[62] R.J. Leeman-Neill, Q. Cai, S.C. Joyce, S.M. Thomas, N.E. Bhola, D.B. Neill, J.L. Arbiser, J.R. Grandis, Honokiol inhibits epidermal growth factor receptor signaling and enhances the antitumor effects of epidermal growth factor receptor inhibitors, Clin. Cancer Res. 16 (2010) 2571–2579.
[63] X. Wang, J.J. Beitler, H. Wang, M.J. Lee, W. Huang, L. Koenig, S. Nannapaneni, A.R. Amin, M. Bonner, H.J. Shin, Z.G. Chen, J.L. Arbiser, D.M. Shin, Honokiol enhances paclitaxel efficacy in multi-drug resistant human cancer model through the induction of apoptosis, PLoS One 9 (2014) e86369.

[64] K. Hall, S. Ran, Regulation of tumor angiogenesis by the local environment, Front. Biosci. (Landmark Ed.) 15 (195–212) (2010) 195–212.
[65] K.T. Hua, W.J. Lee, S.F. Yang, C.K. Chen, M. Hsiao, C.C. Ku, L.H. Wei, M.L. Kuo, M.H. Chien, Vascular endothelial growth factor-C modulates proliferation and chemoresistance in acute myeloid leukemic cells through an endothelin-1-dependent induction of cyclooxygenase-2, Biochim. Biophys. Acta 1843 (2014) 387–397.
[66] J. Wen, A.F. Fu, L.J. Chen, X.J. Xie, G.L. Yang, X.C. Chen, Y.S. Wang, J. Li, P. Chen, M.H. Tang, X.M. Shao, Y. Lu, X. Zhao, Y.Q. Wei, Liposomal honokiol inhibits VEGF-D-induced lymphangiogenesis and metastasis in xenograft tumor model, Int. J. Cancer 124 (2009) 2709–2718.
[67] D.T. Vavilala, V.K. Ponnaluri, R.K. Vadlapatla, D. Pal, A.K. Mitra, M. Mukherji, Honokiol inhibits HIF pathway and hypoxia-induced expression of histone lysine demethylases, Biochem. Biophys. Res. Commun. 422 (2012) 369–374.
[68] K.L. Lan, K.H. Lan, M.L. Sheu, M.Y. Chen, Y.S. Shih, F.C. Hsu, H.M. Wang, R.S. Liu, S.H. Yen, Honokiol inhibits hypoxia-inducible factor-1 pathway, Int. J. Radiat. Biol. 87 (2011) 579–590.
[69] V.K. Ponnaluri, D.T. Vavilala, S. Prakash, M. Mukherji, Hypoxia mediated expression of stem cell markers in VHL-associated hemangioblastomas, Biochem. Biophys. Res. Commun. 438 (2013) 71–77.
[70] A.J. Dannenberg, K. Subbaramaiah, Targeting cyclooxygenase-2 in human neoplasia: Rationale and promise, Cancer Cell 4 (2003) 431–436.
[71] S.H. Liu, C.C. Shen, Y.C. Yi, J.J. Tsai, C.C. Wang, J.T. Chueh, K.L. Lin, T.C. Lee, H.C. Pan, M.L. Sheu, Honokiol inhibits gastric tumourigenesis by activation of 15-lipoxygenase-1 and consequent inhibition of peroxisome proliferator-activated receptor-gamma and COX-2-dependent signals, Br. J. Pharmacol. 160 (2010) 1963–1972.
[72] T.H. Tsai, C.J. Chou, F.C. Cheng, C.F. Chen, Pharmacokinetics of honokiol after intravenous administration in rats assessed using high-performance liquid chromatography, J. Chromatogr. B Biomed. Appl. 655 (1994) 41–45.
[73] F. Chen, T. Wang, Y.F. Wu, Y. Gu, X.L. Xu, S. Zheng, X. Hu, Honokiol: A potent chemotherapy candidate for human colorectal carcinoma, World J. Gastroenterol. 10 (2004) 3459–3463.
[74] W.J. Su, X. Huang, E. Qin, L. Jiang, P. Ren, Pharmacokinetics of honokiol in rat after oral administration of cortex of Magnolia officinalis and its compound preparation Houpu Sanwu decoction, Zhong Yao Cai 31 (2008) 255–258.
[75] X. Wu, X. Chen, Z. Hu, High-performance liquid chromatographic method for simultaneous determination of honokiol and magnolol in rat plasma, Talanta 59 (2003) 115–121.
[76] M. Bohmdorfer, A. Maier-Salamon, B. Taferner, G. Reznicek, T. Thalhammer, S. Hering, A. Hufner, W. Schuhly, W. Jager, In vitro metabolism and disposition of honokiol in rat and human livers, J. Pharm. Sci. 100 (2011) 3506–3516.
[77] X.H. Wang, L.L. Cai, X.Y. Zhang, L.Y. Deng, H. Zheng, C.Y. Deng, J.L. Wen, X. Zhao, Y.Q. Wei, L.J. Chen, Improved solubility and pharmacokinetics of PEGylated liposomal honokiol and human plasma protein binding ability of honokiol, Int. J. Pharm. 410 (2011) 169–174.
[78] Q.Q. Jiang, L.Y. Fan, G.L. Yang, W.H. Guo, W.L. Hou, L.J. Chen, Y.Q. Wei, Improved therapeutic effectiveness by combining liposomal honokiol with cisplatin in lung cancer model, BMC Cancer 8 (2008) 242.

CHAPTER TEN

Effects of Tea Catechins on Cancer Signaling Pathways

Chung S. Yang*,†,1, Hong Wang*, Jayson X. Chen*, Jinsong Zhang†
*Department of Chemical Biology, Ernest Mario School of Pharmacy, Rutgers, The State University of New Jersey, Piscataway, New Jersey, USA
†International Joint Research Laboratory of Tea Chemistry and Health Effects, Anhui Agricultural University, Hefei, PR China
[1]Corresponding author: e-mail address: csyang@pharmacy.rutgers.edu

Contents

Abstract

The inhibitory activities of tea catechins against carcinogenesis and cancer cell growth have been demonstrated in a large number of laboratory studies. Many mechanisms for

The Enzymes, Volume 36
ISSN 1874-6047
http://dx.doi.org/10.1016/B978-0-12-802215-3.00010-0

modulating cancer signaling and metabolic pathways have been proposed based on numerous studies in cell lines with (−)-epigallocatechin-3-gallate, the most abundant and active tea catechin. Nevertheless, the molecular basis for the proposed mechanisms and whether these mechanisms indeed contribute to the anticancer activities *in vivo* are not clearly known. This chapter reviews the basic redox properties of tea catechins, their binding to key enzymes and signal transduction proteins, and other mechanisms that lead to suppression of cell proliferation, increased apoptosis, and inhibition of angiogenesis. More weight is put on studies *in vivo* over experiments *in vitro*. It also discusses key issues involved in extrapolating results from cell line studies to mechanistic insights *in vivo*.

1. INTRODUCTION

Tea, made from the leaves of the plant *Camellia sinensis*, is a widely consumed beverage worldwide. In the past 25 years, the cancer preventive and other anticancer activities of tea preparations have been demonstrated in a variety of animal models in different laboratories (reviewed in Refs. [1–4]). Numerous studies with cell lines have also been carried out in an attempt to understand the mechanisms of the anticancer actions of tea constituents. Because green tea is more active than black tea, it has received more attention. Green tea, consumed mainly in China and Japan, is produced by steaming or pan-frying tea leaves, which inactivates enzymes, and the drying process stabilizes the constituents. When green tea is brewed with hot water, approximately one-third of the solid materials are water-extractable, of which about a third (by dry weight) are catechins. (−)-Epigallocatechin-3-gallate (EGCG), (−)-epigallocatechin (EGC), (−)-epicatechin-3-gallate (ECG), and (−)-epicatechin (EC) are the major polyphenols in green tea (structures shown in Fig. 10.1). They are collectively known as tea catechins. Black tea, the major form of tea consumed in Western countries, India, and other parts of the world, is produced by crushing the tea leaves to promote enzymatic oxidation and subsequent polymerization of most of the catechins to form oligomeric polyphenols (theaflavins) and polymeric polyphenols (thearubigins) [5]. These higher molecular weight black tea polyphenols have very low or no bioavailability, which may explain the lower cancer preventive activity of black tea than green tea. Green and black tea also contain 2–5% caffeine in the water-extractable materials, and caffeine has also been shown to be an active tea constituent in the prevention of skin and lung cancers in animal models [2].

The inhibitory activities of tea catechins against carcinogenesis and the growth of cancer cells have been demonstrated in different experimental

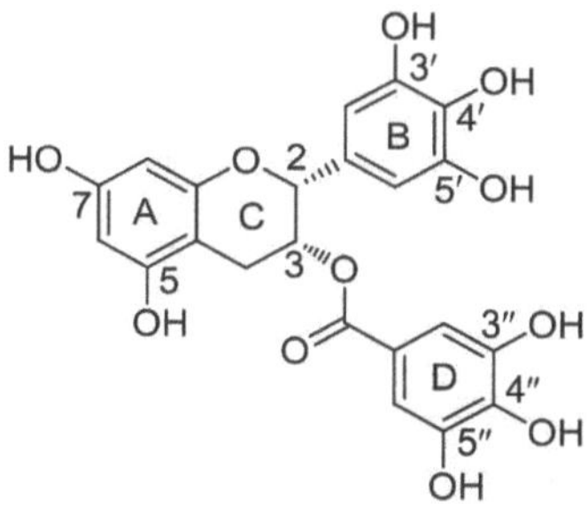

Figure 10.1 Structures of (−)-epigallocatechin-3-gallate (EGCG) and other tea catechins. For (−)-epicatechin-3-gallate (ECG), the "OH" at 3′-position is replaced by "H"; for (−)-epigallocatechin (EGC), the "gallate" at 3-position is replaced by "OH"; and for (−)-epicatechin (EC), the "OH" at 3′-position is replaced by "H" and "gallate" at 3-position is replaced by "OH."

systems [1–4]. Mechanisms of action of tea catechins, especially EGCG, the most abundant and active form of catechin, have been extensively investigated in cell culture systems [2,6]. This is a very active area of research. A PubMed search in June 2014 under the title "Tea and Cancer" yielded 3621 citations (1962–2014), while "EGCG and cancer signaling" yielded 300 publications (1997–2014). EGCG has been shown to affect a variety of cancer signaling pathways. While it is surprising that a single molecule such as EGCG can have such diverse activities, it is also unclear whether these proposed mechanisms are indeed involved in the inhibition of carcinogenesis or tumor growth in animal models and humans.

This chapter reviews the chemical and biochemical properties of tea catechins and their effects on cancer signaling and metabolism. Most of the information is about EGCG, the most frequently studied catechin. Some results from our own laboratory are discussed in more detail to serve as examples and to illustrate the challenges in using results from *in vitro* studies to gain insight into the action of tea catechins and other natural products in animal models and humans.

2. CHEMISTRY, BIOAVAILABILITY, AND BIOTRANSFORMATION OF TEA CATECHINS

2.1. Chemistry

The major catechins in green tea as shown in Fig. 10.1 are characterized by the dihydroxyl or trihydroxyl substitutions on the B ring and the *m*-5,7-dihydroxyl substitutions on the A ring [5]. The B ring appears to be the principal site of antioxidant reactions, and the antioxidant activity is further increased by the trihydroxyl structure in the D ring (gallate) in EGCG and ECG [5]. The polyphenolic structure allows electron delocalization, conferring the ability to quench free radicals. Tea catechin preparations have been shown to reduce reactive oxygen species (ROS) such as superoxide radical, singlet oxygen, hydroxyl radical, peroxyl radical, nitric oxide, nitrogen dioxide, and peroxynitrite [5]. Among tea catechins, EGCG is most effective in reacting with the majority of ROS. Tea polyphenols are also strong chelators of metal ions such that the chelation of free metal ions prevents the formation of ROS.

The vicinal dihydroxy or trihydroxy structures not only contribute to the antioxidant activity of tea catechins but also increase the susceptibility of these compounds to air oxidation under alkaline or neutral pH, especially in the presence of trace amounts of cuprous or ferric ion. Auto-oxidation of EGCG generates superoxide anion and hydrogen peroxide and leads to the formation of catechin dimers, such as theasinensins, which are unstable and yield products yet to be identified [7]. We propose that this is due to superoxide anion-mediated chain reactions outside of the cells, because EGCG can be stabilized by the addition of superoxide dismutase (SOD) [7]. These reactions occur under cell culture conditions, and the ROS generated by EGCG can induce many cellular changes and may be responsible for many of the activities reported in the literature [1,2,7].

2.2. Bioavailability

The polyphenolic structure of tea polyphenols makes them good donors for hydrogen bonding. Hydrogen bonding of water molecules to EGCG forms a large hydration shell, which reduces the absorption of EGCG. The bioavailability of tea polyphenols follows the Lipinski's Rule of Five [8] and is dependent on the molecular size, apparent size (due to the formation of a hydration shell), and polarity. For example, the bioavailabilities of EC (molecular weight 290 and 5 phenolic groups) are much higher than EGCG (molecular weight 458 and 8 phenolic groups). In humans, following the oral administration of the equivalent of two or three cups of decaffeinated green tea, the peak plasma levels of EGCG (including the conjugated forms) were usually 0.2–0.3 μM [9]. The plasma concentrations of EGC (molecular weight 306 and 6 phenolic groups) were higher, even though this green tea preparation contains less EGC than EGCG. With high pharmacological oral doses of EGCG, peak plasma concentrations of 2–9 and 7.5 μM were observed in mice and humans, respectively [9]. EGCG and other catechins are thought to enter cells through passive diffusion. However, the involvement of transporters, such as organic anion-transporting peptides 1A2 and 1B3 [10], has been suggested. Active efflux has been shown to limit the bioavailabilities of many polyphenolic compounds, including catechins. The multidrug resistance-associated protein 2, located on the apical surface of the intestine and liver, mediates the transport of some polyphenolic compounds to the lumen and bile, respectively [11]. EGCG and its metabolites are predominantly effluxed from the enterocytes into the intestinal lumen or from the liver to the bile and excreted in the feces, with little or none of these compounds excreted in the urine [5].

2.3. Biotransformation

EGCG and other tea catechins undergo extensive biotransformations (reviewed in Ref. [9]). Because of the catechol structure, EGCG and other catechins are readily methylated by catechol-*O*-methyltransferase as a detoxification mechanism. In addition, catechins are glucuronidated by UDP-glucuronosyltransferases and sulfated by sulfotransferases. Multiple methylation and conjugation reactions can occur on the same molecule [5]. Tea catechins can also be degraded in the intestinal tract by microorganisms. After ingestion of tea catechins by humans, ring fission metabolites 5-(3′,4′,5′-trihydroxyphenyl)-γ-valerolactone (M4), 5-(3′,4′-dihydroxyphenyl)-γ-valerolactone (M6), and 5-(3′,5′-dihydroxyphenyl)-γ-valerolactone (M6′) have been observed in urine and plasma samples [12]. These compounds can undergo further degradation to phenylacetic and phenylpropionic acids in the intestine and are excreted in feces and urine.

3. INHIBITION OF TUMORIGENESIS BY TEA CATECHINS IN ANIMAL MODELS AND POSSIBLE MECHANISMS

Tea extracts have been demonstrated to inhibit tumorigenesis in many animal models, including those for cancers of the oral cavity, esophagus, stomach, small intestine, colon, liver, pancreas, lung, bladder, skin, prostate, and mammary glands. Most of the studies were conducted with green tea or tea catechin preparations, and some were conducted with pure EGCG, administered through drinking water or diet. Some examples of these studies are as follows.

3.1. Inhibition of Tumorigenesis in the Digestive Tract

The systemic bioavailability of gallated catechins (EGCG and ECG) is a limiting factor for their effectiveness against tumorigenesis in the internal organs. The epithelial cells in the digestive tract have the advantage of being directly exposed to the catechins that are ingested orally. Inhibitory effects of tea catechins against tumorigenesis in chemically induced and genetic animal cancer model in the oral cavity, esophagus, stomach, small intestine, and colon have been shown in more than 30 studies. For example, we showed that administration of EGCG at 0.02–0.32% in drinking water dose-dependently inhibited small intestinal tumorigenesis in $Apc^{Min/+}$ mice, while caffeine did not have an inhibitory effect [13]. The inhibition was associated with increased levels of E-cadherin on the plasma membrane, as well as decreased levels of nuclear β-catenin, c-Myc, phospho-AKT, and phospho-ERK1/2 in the tumors [13]. Administration of green tea extracts (0.6% in drinking fluid) also inhibited the formation of azoxymethane (AOM)-induced aberrant crypt foci (ACF) in CF-1 mice on a high-fat diet [14]. Treatment of rats with 0.24% of Polyphenon (PPE, a standard tea catechin preparation containing 65% EGCG and other tea catechins) in the diet for 8 weeks decreased the total number of ACF per rat by 36.9%. In ACF with high-grade dysplasia, the inhibitory activity of PPE was associated with decreased levels of nuclear β-catenin and cyclin D1, and increased retinoid X receptor-α staining [15]. Recently, Shimizu *et al.* [16] demonstrated the inhibition of AOM-induced ACF formation in male C57BL/KsJ-*db/db* mice by EGCG (0.01% and 0.1% in drinking water) involving suppression of insulin-like growth factor 1 (IGF1) signaling. The elevated levels of IGF1 receptor (IGF1R), phospho-IGF1R, phospho-GSK3β, and β-catenin in the colonic mucosa were decreased by treatment with EGCG; also decreased were the plasma levels of IGF1, insulin, triglyceride, cholesterol, and leptin [16].

3.2. Inhibition of Lung Tumorigenesis

The inhibitory effects of tea catechins against lung tumorigenesis have been demonstrated in at least 20 studies using chemically induced and genetic rodent models [1]. Administration of EGCG or EGC significantly decreased lung tumorigenesis in rats, mice, or hamsters [1,17]. Oral administration of 0.5% PPE or 0.044% caffeine in the drinking water, to tumor-bearing A/J mice [induced by a single dose of (4-methylnitrosamino)-1-(3-pyridyl)-1-butanone 20 weeks earlier] for 32 weeks, inhibited the progression of lung adenomas to adenocarcinomas [17]. Immunohistochemical analysis showed that PPE and caffeine treatment inhibited cell proliferation, enhanced apoptosis, and decreased levels of c-Jun and phospho-ERK1/2 in adenocarcinomas. In normal lung tissues, neither agent had a significant effect on cell proliferation or apoptosis, suggesting that the action is selective against tumor tissues. These results demonstrate the broad inhibitory activity of tea catechins against lung carcinogenesis, but the mechanisms remain to be further elucidated.

3.3. Inhibition of Prostate Carcinogenesis

Administration of a green tea polyphenol infusion (0.1% in drinking water) to transgenic adenocarcinoma of the mouse prostate (TRAMP) mice for 24 weeks markedly inhibited prostate cancer development and distant site metastases [18,19]. The inhibition was associated with decreased cell proliferation, increased apoptosis, decreased IGF1 level, and restored IGF-binding protein 3 (IGFBP3) levels in both serum and the dorsolateral prostate [18,19]. This modulation of IGF1 and IGFBP3 levels was associated with reduced levels of phosphatidylinositol 3-kinase (PI3K) as well as phosphorylated forms of AKT and ERK1/2. The green tea polyphenol treatment also significantly decreased levels of angiogenic and metastatic markers, such as vascular endothelial growth factor A (VEGFA), matrix metalloproteinase (MMP)2, and MMP9. These results suggest that the inhibition of the IGF1 signaling, VEGFA, and MMPs contributes to the cancer preventive activity of green tea polyphenols.

3.4. Human Studies

In contrast to the strong evidence for the cancer preventive activity of tea constituents in animal models, results from epidemiological studies have not been consistent concerning the cancer preventive effect of tea consumption in humans [1,2]. There are only few intervention studies showing promising cancer preventive effects of tea preparations [1,2]. As for therapeutic applications, only a few clinical trials demonstrated the usefulness of EGCG. For example, in a phase 2 trial in patients with early chronic lymphocytic leukemia, oral doses of PPE (2000 mg twice daily) caused durable declines in the absolute lymphocyte count and/or lymphadenopathy in the majority of patients [20]. Side effects observed in this study include transaminitis, abdominal pain, and diarrhea.

The difference between the results from most animal and human studies is likely to be due to (1) the lower quantities of human tea consumption as compared to the doses used in animal studies and the relatively weak cancer preventive or therapeutic effects of tea catechins in humans, and (2) confounding factors in epidemiological studies; whereas in animal studies, the conditions are controlled to maximize the opportunity to detect a cancer prevention effect. The results of human studies on tea and cancers may become more understandable after considering the quantities of tea consumption and correction for smoking and other interfering factors.

4. BIOCHEMICAL ACTIVITIES OF TEA CATECHINS

To understand the mechanisms by which EGCG and other catechins modulate cancer signaling, it is important to consider the basic biochemical activities of these compounds. The primary actions of catechins are due to their redox and physical binding activities. Using EGCG as a prototype compound, these actions are depicted in Fig. 10.2A and discussed below. The effects of EGCG at the receptor and cellular levels are shown in Fig. 10.2B.

4.1. Antioxidant and Pro-oxidative Activities *In Vitro* and *In Vivo*

Catechins are well recognized as antioxidants, but they can also be pro-oxidants and generate ROS. ROS can alter the functions of cellular proteins,

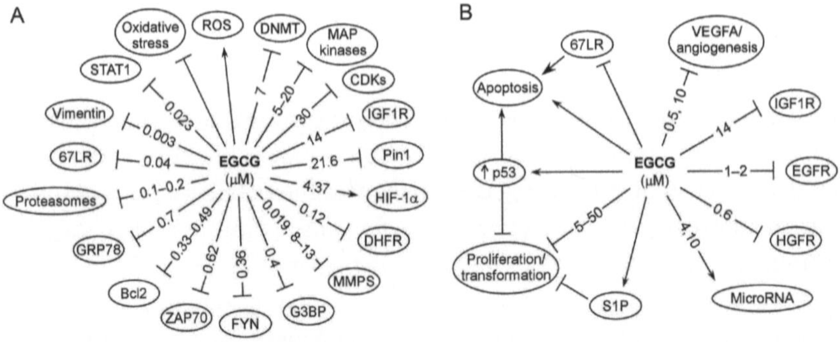

Figure 10.2 Possible targets for the cancer preventive activity of EGCG. Some of these are direct binding targets; others are affected indirectly. The reported effective concentrations, in IC_{50}, K_i (inhibition constant), or K_d (dissociation constant), are shown in μM. All these are from studies *in vitro*. When two values are given, the first value is from cell-free systems and the second value is from studies in cell lines.

lipid, and nucleic acids and lead to different diseases [21]. Oxidative damages to DNA cause mutation and genomic instability, which are major contributing factors in the initiation, promotion, and progression of carcinogenesis [22]. However, there are also suggestions that decreasing the amounts of ROS promotes tumor progression [23,24]. Therefore, the antioxidant and pro-oxidative activities of catechins have the potential to affect cancer signaling, depending on the bioavailability of catechins and context of the cellular environment. Although the antioxidant activity of tea catechins is well established *in vitro* [5], such activity *in vivo* is only observed under circumstances when the animals are under oxidative stress. For example, EGCG administration has been found to decrease the levels of lipid peroxidation and protein carbonylation in old rats, but not in young rats [25]. In animal models for carcinogenesis, ROS are induced by the treatment with carcinogens, and EGCG has been demonstrated to reduce the formation of 8-hydroxydeoxyguanosine (8-oxo-dG), a well-established marker for oxidative DNA damage that can mispair to induce mutations [26]. As endogenously formed ROS are important in promoting carcinogenesis, tea polyphenols may have important roles in quenching these species at different stages of carcinogenesis. In human studies, administration of green tea to smokers for 4 weeks has been shown to significantly reduce the number of 8-oxo-dG-positive cells [27]. Such antioxidant actions of tea catechins may be important in the prevention of carcinogenesis.

Tea catechins can be auto-oxidized to generate ROS in cell culture medium and cause cell death [7,28]. After entering the cells, EGCG may also induce the production of ROS in the mitochondria. In our studies, oral administration of EGCG to mice bearing human lung cancer H1299 cell xenograft tumors inhibited tumor growth, enhanced tumor cell apoptosis, and produced ROS in the tumor cells [29]. The observed ROS production in tumor cells is probably due to the lack of sufficient antioxidant enzymes in H1299 cells. It remains to be demonstrated whether the production of ROS is responsible for the induction of apoptosis *in vivo*. At modest doses (e.g., 0.5% EGCG in the diet), although increased levels of 8-oxo-dG and phosphorylated histone 2A variant X were seen in xenograft tumors, 8-oxo-dG production and toxicity were not observed in the liver, kidney, and other organs of the host mice [29]. At high doses (e.g., 750 mg/kg, i.g.), however, hepatotoxicity and ROS were observed [30]. These toxic responses are probably similar to the reported liver toxicity in individuals who took excessive amounts of tea extracts in dietary supplements used for the purpose of weight reduction [31,32].

Cellular ROS may also activate the nuclear factor erythroid 2-related factor 2 (Nrf2)-mediated signaling pathways to induce cytoprotective enzymes [33]. For example, oral gavage of EGCG (200 mg per kg) to C57BL/6J mice upregulated gene expression of γ-glutamyltransferase, glutamate cysteine ligase, and hemoxygenase 1 in the liver and colon, which were most likely mediated by the activation of Nrf2 [34]. Similarly, human volunteers supplemented with 800 mg PPE per day for 4 weeks increased glutathione *S*-transferase P activity in lymphocytes [35]. In an intervention study in a high aflatoxin exposure area in China, supplementation with 500 or 1000 mg green tea polyphenols per day for 3 months increased the median urinary aflatoxin B1-mercapturic acid levels by more than 10-fold compared to baseline [36]. This result is likely due to the induction of glutathione *S*-transferase by EGCG. It appears that levels of ROS produced by moderate doses of tea polyphenols activate Nrf2 to reduce oxidative stress, but high doses of tea polyphenols can produce high levels of ROS, which induce toxicity [31]. Thus, the biological effect of EGCG depends on the dose used and the context of the biological systems.

4.2. High-Affinity Binding Proteins as Targets of EGCG

EGCG is known to bind to a variety of proteins with rather high affinities, through hydrogen bonding and binding to the hydrophobic pockets of proteins. For example, a recent report on the interaction between EGCG and bovine serum albumin (BSA), studied with fluorescence lifetime and intensity, has suggested that EGCG binds to BSA with an apparent binding constant of about $4.6 \times 10^{-5}\ M^{-1}$ at pH 7.4 [37]. EGCG binds with high affinity to the hydrophobic pocket between subdomains IIA and IIIA of BSA. The BSA molecule also undergoes a binding-dependent conformational change. The multiple phenolic groups of EGCG can serve as hydrogen bond donors to many proteins and nucleic acids. In our previous work with molecular modeling, the binding of EGCG to DNA methyltransferase (DNMT) 1 has been proposed to involve five hydrogen bonds [38]. In an earlier study, using NMR spectroscopy, EGCG was demonstrated to directly bind to the BH3 pocket of antiapoptotic Bcl_2 proteins—with inhibition constant (K_i) of 0.33–0.49 μM [39]. However, higher EGCG concentrations (by 2 orders of magnitude) were needed to induce apoptosis.

Using an EGCG–Sepharose 4B column and 2D gel electrophoresis, Dong *et al.* identified vimentin [40], IGF1R [41], FYN [42], glucose-regulated protein 78 kDa (GRP78) [43], ZAP70 [44], and Ras-GTPase-activating protein SH3 domain-binding protein 1 (G3BP1) [45] as high-affinity EGCG-binding proteins (Fig. 10.2). Functionally, for example, EGCG was shown to competitively bind to the ATP-binding site of IGF1R and inhibit tyrosine phosphorylation [41]. A subsequent X-ray crystallography study by Dong's group demonstrated the binding of EGCG to both the WW and PPIase domains of peptidyl prolyl *cis/trans* isomerase (Pin1) [46]. The direct binding of EGCG with Pin1 was confirmed and the binding inhibited Pin1 PPIase activity. This inhibition could have important biological consequences because Pin1 is required for full activation of AP-1, NFκB, β-catenin, and other signaling pathways. Biochemical studies showed a dissociation constant of 21.6 μM for the binding of EGCG to Pin1. EGCG was shown to suppress the proliferation of cells expressing Pin1 and tumor growth in a xenograft mouse model. The binding of EGCG with Arg17 in the WW domain prevented its binding with c-Jun, a well-known Pin1 substrate. EGCG treatment also decreased cyclin D1 level and 12-*O*-tetradecanoylphorbol-13-acetate-induced AP-1 or NFκB activity in cells expressing Pin1 [46].

In a recent study with surface plasmon resonance (SPR), molecular modeling, and site-directed mutagenesis, EGCG (and ECG) were found to tightly bind to signal transduction activator of transcription 1 (STAT1) with K_d of 23 nM in MDA-MB-231 breast cancer cells [47]. The binding involved at least three hydroxyl groups of the B ring and one hydroxyl group of the D ring of EGCG. Site-directed mutagenesis of STAT1 with H568A eliminated the high-affinity binding to EGCG. The tight binding of EGCG to STAT1 blocked its phosphorylation by Janus kinase 2 (JAK2). This could be a mechanism by which EGCG inhibits STAT1 activation and related actions on cytokines and growth factors. EGCG was also suggested to inhibit the JAK/STAT3 signaling and lead to Fas/CD95-mediated apoptosis of head and neck squamous carcinoma cells [48]. Previously, Tachibana *et al.* used SPR to study the binding of EGCG to the 67 kDa laminin receptor (67LR) [49] and this led a series of interesting studies which will be discussed in a later section.

All of the aforementioned proteins were demonstrated to be important for the inhibitory activity of EGCG in some cell lines. However, much higher EGCG concentrations (than the K_d values) were needed to elicit a cellular response. For example, vimentin bound to EGCG with a K_d of 3.3 nM, and functional studies showed that EGCG inhibited the phosphorylation of vimentin with $IC_{50} = 17$ μM. The reason for the difference in effective concentrations is not fully understood. It is possible that EGCG binds nonspecifically to proteins and other macromolecules in the cells and therefore elevates the effective concentrations of EGCG at the targets. The discovery of the aforementioned high-affinity EGCG-binding proteins is important, and the direct involvement of these proteins in the action of EGCG in animal models and humans remains to be investigated.

4.3. Inhibition of Enzyme Activities

Tea polyphenols have been shown to bind and inhibit the activities of a variety of enzymes. In other studies, the inhibition of enzyme activity has been observed in cultured cells and the inhibition could be due to either direct binding or indirect actions. We previously observed that EGCG, at concentrations of 5–20 μM, inhibited the phosphorylation of JNK (Jun N-terminal kinase), c-Jun, MEK1/2, ERK1/2, and ELK1 in JB6 epidermal cell lines [50]. This inhibition was associated with the inhibition of AP1 transcriptional activity or cell transformation. Additional studies with *in vitro* kinase assays suggested that EGCG inhibited MEK1/2 phosphorylation by decreasing its association with the kinase RAF1 [51]. Moreover, EGCG seemed to inhibit the phosphorylation of ELK1 by competing with the binding site for ERK1/2 [52]. Recent studies also suggest that EGCG inhibited the phosphorylation of ERK1/2 and AKT in Epstein–Barr virus (EBV)-positive cells, and this could block the constitutive lytic infection of EBV at the gene transcription and translation levels [53] There were also studies showing that EGCG activated ERK1/2 and other MAP kinases through the generation of ROS, but these results could be *in vitro* artifacts. In lung carcinogenesis models, EGCG and a green tea polyphenol preparation have been shown to inhibit the phosphorylation of c-Jun and ERK1/2 [17].

EGCG has also been reported to inhibit the chymotryptic activity of 20S proteasomes [54], which is a key step in the degradation of many signaling proteins. The difference in the effective concentrations of EGCG in cell-free systems (IC_{50}=0.09–0.2 μM) and in cell lines (IC_{50}=1–10 μM) [54] is probably due to nonspecific binding to different macromolecules. MMPs are secreted by tumor cells during cancer cell invasion and metastasis. EGCG and other catechins have been shown to inhibit the activity of secreted MMP2 and MMP9 with IC_{50} values of 8–13 μM [55,56]. In addition, EGCG could also increase the expression of the tissue inhibitor of MMPs (TIMP1 and TIMP2) at lower concentrations (~1 μM) [56]. Recent studies demonstrated that EGCG induced TIMP3 by epigenetic mechanisms in MCF7 and MDA-MB 231 breast cancer cells, involving decreased protein levels of the enhancer of zeste homolog 2 and class 1 histone deacetylases (HDACs) [57]. These activities may contribute to the reported inhibition of metastasis and invasion following treatment of tumor-bearing mice with green tea or EGCG [58]; however, additional *in vivo* studies are needed to verify this mechanism.

We reported previously that EGCG inhibited DNMT activity ($K_i = 7\ \mu M$) from KYSE 510 human esophageal cancer cells, and this resulted in the demethylation and reactivation of the hypermethylated promoters of the tumor suppressor gene *INK4A*, *retinoic acid receptor β*, as well as the DNA repair genes, *MLH1* and *methylguanine methyltransferase* [38]. Reactivation of some of these genes was also observed in HT29 colon and PC3 prostate cancer cells. Tea catechins also reactivated GSTP1 in human prostate cancer cells by causing promoter hypomethylation and chromatin remodeling [59]. However, the induction of promoter demethylation by EGCG was not observed in some other hypermethylated genes. As will be discussed in a later section, EGCG has been suggested to induce other epigenetic events.

EGCG has also been reported to inhibit dihydrofolate reductase [60], glucose-6-phosphate dehydrogenase [61], glyceraldehyde-3-phosphate dehydrogenase [62], and carbonyl reductase 1 [63]. These results on enzyme inhibition are interesting biochemically. Cancer metabolism is an active area for research. Additional studies in this area are needed, especially regarding specificity to target enzymes in cancer cells versus normal cells.

5. MODULATING SIGNALING PATHWAYS AND CELL FUNCTIONS

Most of the information in this area was derived from studies in cell lines, mainly with EGCG. Some of the activities are summarized in Fig. 10.2B. When a similar action was also observed *in vivo*, then the mechanism is more convincing. Some studies have also included other catechins; their activities generally follow the ranking order of EGCG > ECG > EGC > EC.

5.1. Inhibition of Receptor Tyrosine Kinases and Other Receptors

Tea catechins have been shown to affect many receptor-related activities, and their inhibitory actions against receptor tyrosine kinases (RTKs) have been reviewed recently by Larsen *et al.* [64] and Shimizu *et al.* [65]. All members of the RTK family, including epidermal growth factor (EGFR), IGF1R, hepatocyte growth factor receptor (HGFR or c-Met), and vascular endothelial growth factor receptor (VEGFR), consist of an extracellular ligand-binding domain, single membrane-spanning region, and a cytoplasmic protein tyrosine kinase domain. The major signaling pathways activated by RTKs are the Ras/ERK and the PI3K/AKT pathways. Some of the actions, using EGFR as an example, are depicted in Fig. 10.3. Members of the EGFR family are frequently overexpressed in human cancers and

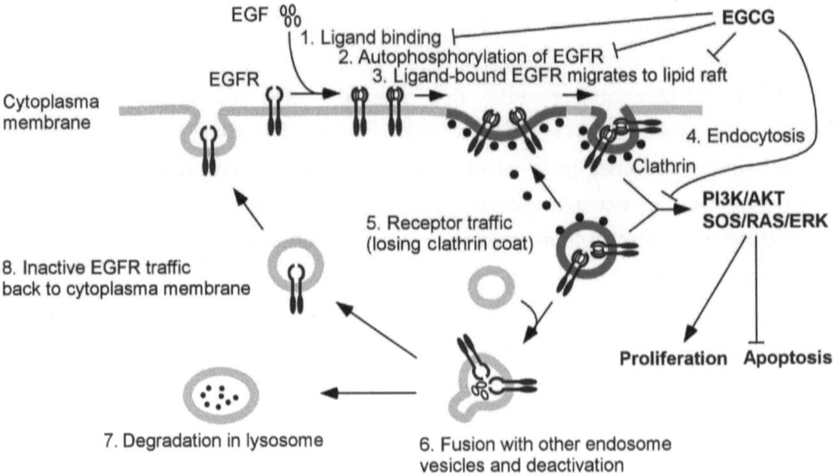

Figure 10.3 Possible mechanisms of inhibition of receptor tyrosine kinases by EGCG. Epidermal growth factor receptor (EGFR) is used as an example to illustrate the multiple intracellular processes in signaling. (1) Upon epidermal growth factor (EGF) binding, (2) the receptor undergoes autophosphorylation and conformational changes, transforming the EGFR to the active form on the surface of cytoplasmic membrane. (3) Once activated, EGFR is in a functional membrane with unique lipid composition or mobility (often referred to as lipid raft). (4) Such a functional lipid unit of EGFR, mediated by clathrin, internalizes and the active EGFR signaling is transduced by the activation of downstream PI3K/AKT and SOS/RAS/ERK pathways. (5) Clathrin-coated internalized vesicles are decoated, and EGF is disassociated from the active EGFR. (6) The decoated vesicles are fused with other intracellular vesicles. (7) When fused with lysosomes, EGFR is degraded. (8) When fused with vesicles from Golgi, EGFR can be recycled back to cytoplasmic membrane. The signaling transduction mediated by different membrane units or subcellular components is also found in other receptor tyrosine kinases such as IFGR, HGFR, and VEGFR. EGCG has been reported to inhibit this signaling pathway by interfering with the binding of EGF to EGFR, inhibiting EGFR kinase activity, altering lipid organization in the plasma membrane (lipid raft), and inducing EGFR internalization without activation as discussed in the text.

are associated with poor prognosis [66]. Many studies have demonstrated the inhibitory effects of EGCG on the EGFR signaling pathways [7,67–70]. The different mechanisms that have been proposed for the inhibition of EGFR by EGCG include (1) interfering with the binding of EGF to EGFR and inhibiting EGFR tyrosine kinase activity [67], (2) altering lipid organization in the plasma membrane (lipid rafts) and inhibiting EGF binding to EGFR [69], and (3) inducing EGFR internalization without activation [70]. The synergistic action of EGCG and erlotinib, an EGFR tyrosine kinase inhibitor, against head and neck cancer cell growth has been reported [71].

IGF1R activation by IGF1 can induce cell proliferation, cell survival, transformation, metastasis, and angiogenesis as well as inhibit apoptosis in different cancer cell lines [72]. IGF/IGF1R axis has been reported to be targets of EGCG in human colon and hepatocellular carcinoma cells [73,74]. EGCG also inhibits IGF1R phosphorylation and increases expression of transforming growth factor-β2 in human colon cancer SW837 cells [73], which is consistent with our observations in H-Ras-transformed human bronchial epithelial 21BES cells [75]. Direct binding of EGCG to IGF1R as proposed by Li *et al.* [41], as discussed previously, is a likely mechanism. These results from cell line studies are consistent with those from animal studies showing that orally administrated EGCG and other tea catechins inhibited the IGF/IGF1R axis in a colon carcinogenesis model in *db/db* obese mice [16] and TRAMP mice [18].

Inhibition of EGFR signaling has also been shown to decrease the production of VEGFA in cancer cells [76]. EGCG (0.5–10 μM) disrupted VEGFA-induced VEGFR2 dimerization in human umbilical vein endothelial cells [77] and inhibited growth and activation of VEGF/VEGFR axis in human colorectal cancer cells [78]. We have previously observed the down-regulation of VEGFA expression and suppression of angiogenesis by treatment with green tea (0.6% green tea solid in drinking fluid) in a lung tumorigenesis model [79]. In a murine gastric tumor model, EGCG suppressed VEGFA protein expression and tumor microvessel density [80].

Deregulation of HGFR pathway occurs in several types of human cancers and can lead to increased tumorigenesis and metastasis [81]. HGFR and HGF play key roles in epithelial–mesenchymal transition, which is associated with tumor invasion [82]. It has been shown in MDA-MB-231 cells that the HGF-induced phosphorylation of HGFR and AKT is completely blocked by 0.6 μM EGCG, and that cell invasion is significantly decreased by 5 μM EGCG [83]. Larsen *et al.* have provided evidence for the binding of EGCG to the ATP-binding site of HGFR [84]. In a study with FaDu hypopharyngeal carcinoma cells, 1 μM EGCG prevented HGF-induced motility in an *in vitro* wound healing assay [85]. In a series of nonsmall cell lung cancer cell lines, EGCG was also found to be a potent inhibitor of cell proliferation and appeared to be more effective against HGFR than against EGFR [86].

EGCG has also been suggested to transcriptionally target sphingosine-1-phosphate receptor S1P2 and prevent sphingosine-1-phosphate-mediated signaling in macrophage-differentiated HL-60 promyelocytic leukemia cells [87].

5.2. Effects on 67LR

Binding of EGCG to the 67LR (with a K_d value of 0.04 μM) was first observed by Tachibana *et al.* using an SPR assay [49]. Expression of the metastases-associated 67LR increased the responsiveness of MCF7 cells to low micromolar concentrations of EGCG [49]. RNA interference (RNAi)-mediated silencing of 67LR abrogated EGCG-induced apoptosis in multiple myeloma (MM) cells [88]. Further studies by this research group also demonstrated the critical role of 67LR in mediating anti-inflammation action of EGCG (1 μM) in macrophages [89]. Anti-67LR antibody treatment or RNAi-mediated silencing of 67LR resulted in the abrogation of the inhibitory action of EGCG on lipopolysaccharide-induced activation of TLR4 and downstream signaling of inflammation. Recent work by Kumazoe *et al.* [90] showed that the activation of 67LR by EGCG in primary MM cells and MM cell lines (U266, ARH-77, and RPMI 8226) resulted in elevated levels of cGMP, which initiated apoptosis through the activation of PKCδ and acid sphingomyelinase in a novel death pathway. However, EGCG alone was not very effective in killing MM U266 cells (IC_{50} of 23.2 μM), because these cells overexpressed phosphodiesterase 5 (PDE5), a negative regulator of cGMP. When a PDE5-selective inhibitor vardenafil was also added to cultured cells, it synergized with EGCG to reduce the IC_{50} of EGCG to 1.4 μM. These authors also described the synergy between EGCG and other PDE5-selective inhibitors (zaprinast, MQZ, and sildenafil), and used shRNA, neutralizing antibodies, and enzyme inhibitors to demonstrate the involvement of this key receptor and related enzymes in the synergistic action. This impressive synergism was also shown in MM MPC-11 cells and MDA-MB-231 breast cancer cells in a xenograft model, as well as *in vitro* in some gastric, pancreatic, and prostate cancer cell lines. Overexpression of both 67LR and PDE5 was observed in these cell lines [90].

5.3. Inhibition of Wnt Signaling

The Wnt signaling involves the nuclear translocation of β-catenin to transcriptionally activate proto-oncogenes such as c-Myc, cyclin D1, and COX-2. Our studies in *Apc*$^{min/+}$ mice suggested that EGCG inhibits Wnt signaling [13]. Treatment of HT29 human colon cancer cells with EGCG (20 μM) decreased nuclear levels of β-catenin as well as cellular levels of c-Myc and cyclin D1 [13]. Our recent studies in colon cancer cell lines also suggested that EGCG induced β-catenin N-terminal phosphorylation at the Ser33/37 residues and promoted its degradation in *Apc* mutated colon cancer cell lines (unpublished results). The EGCG-induced β-catenin phosphorylation and degradation is consistent with a similar observation by Singh *et al.* [91]. They also reported that EGCG downregulated the prostaglandin E2-induced levels of MMP2 and MMP9, which are downstream targets of β-catenin [91]. EGCG was also shown to inhibit the Wnt signaling in hepatoblastoma cells [92]. Interestingly, this was found to be associated with the reexpression of the silenced tumor suppressor gene, secreted frizzled-related protein (SFRP)1, which is known to modulate Wnt signaling.

5.4. Epigenetic Mechanisms

5.4.1 Affecting Epigenetic DNA Methylation and Histone Modification

In addition to the aforementioned epigenetic changes [38,57,59,92], EGCG was reported to decrease the levels of 5-methylcytosine, DNMT activity, and expression levels of DNMT1, DNMT3a, and DNMT3b in human epidermoid carcinoma A431 cells. It also decreased HDAC activity and affected levels of acetylated lysines on histones H3 and H4 [93]. Furthermore, EGCG inhibited the majority of acetyltransferase (HAT) enzymes [94] and inhibit the transcription of hTERT (human telomerase reverse transcriptase), the catalytic subunit of telomerase, through epigenetic mechanisms mediated at least partially through the inhibition of DNMT and HAT activities [95].

5.4.2 Effect on MicroRNA

MicroRNAs are small (about 22 bases) single-stranded, endogenous non-coding RNAs that negatively regulate the translation and/or stability of mRNAs [96]. MicroRNA levels could be altered by EGCG to cause subtle changes in multiple molecular targets and pathways. It has been reported that

EGCG upregulated miR-16 in human hepatocellular carcinoma HepG2 cells, and this led to the downregulation of Bcl2 and induction of apoptosis [97]. In our recent work in both human and mouse lung cancer cells in culture, we found that EGCG specifically upregulated the expression of miR-210, a major microRNA regulated by HIF-1α [98]. The upregulation of miR-210 was found to be correlated with the transiently stabilized HIF-1α in lung cancer cell lines after EGCG treatment. We also demonstrated that EGCG could bind to the oxygen-dependent degradation domain of the hypoxia-response element of HIF-1α promoter and prevented the hydroxylation-dependent ubiquitination and proteasome-mediated degradation of HIF-1α. The *in vivo* relevance of this observation, however, remains to be demonstrated. A recent study also showed that EGCG upregulated miR-16 in breast cancer cell line 4T1 [99]. The miR-16 could be transfected to tumor-associated macrophages (TAMs) via exosomes and inhibited TAM infiltration and M2 microphage polarization. These actions were suggested to be responsible for the observed growth suppression of xenograft tumors from 4T1 cells in BALB/c mice treated with EGCG.

5.5. Other Mechanisms

5.5.1 Modulating p53-Dependent Events

Many studies have demonstrated that EGCG treatment induces p53 expression and p53-dependent apoptosis. Some studies suggested that this was mediated by ROS, which are produced in the auto-oxidation of EGCG [100], while others showed this did not involve changes in intracellular levels of ROS [101]. It was also reported that the increased p53 transcription and acetylation were due to the suppression of class I HACs in LNCaP cell [102]. In another study, EGCG was found to induce the expression of p53 in HepG2 cells and blocked the cell cycle progression at the G1 phase, whereas EGCG induced apoptosis in p53-negative Hep3B cells [103]. The involvement of p53 in the biological activity of EGCG requires additional studies *in vitro* and *in vivo*.

5.5.2 Binding to Lipids

The possibility that EGCG alters lipid organization in the plasma membrane (lipid rafts) and affects protein distribution and receptor functions has been proposed for the inhibition of the functions of EGFR [69], HGFR [104], and 67LR [105]. Although interesting, it remains to be determined whether the effects occur in normal cells, whether EGCG also alters the lipid rafts of cancer cells *in vivo*, and what concentrations of EGCG are required to exert an observable effect *in vivo*.

5.5.3 Binding to Nucleic Acids

Based on the physical binding of EGCG to nucleic acids, it has been suggested that DNA and RNA can also be targets of action of tea catechins [106]. However, the relevance of this proposed binding depends on whether the catechins can bind selectively to specific nucleic acid in the genome of cancer or premalignant cells without affecting normal cells.

6. ISSUES IN EXTRAPOPULATING STUDIES *IN VITRO* TO SITUATIONS *IN VIVO*

In relating observations *in vitro* to molecular events *in vivo*, an important issue is the difference in effective concentrations. In most animal experiments, in which an inhibitory effect of EGCG could be observed, the EGCG levels in the blood and tissues were usually in the submicromolar levels. How do we evaluate the relevance of an experiment using 20–100 μM EGCG in cell cultural studies? To address the relationship between the effective concentrations *in vivo* versus *in vitro*, we measured the EGCG levels in blood and xenograft tumors of H1299 lung cancer cells. At conditions when tumor growth was inhibited approximately 50% by dietary EGCG (0.5%), the EGCG concentrations was 0.52 μM in blood and 0.18 μM in tumor tissues. This value is 2 orders of magnitude lower than the IC_{50} values of EGCG in the inhibition of H1299 cell growth in culture [29]. One possible reason for the observed discrepancy between the cell culture system and the xenograft model is the rather short-term exposure to EGCG in cell culture studies (24 or 48 h) compared to the long-term treatment in animal models. It has been reported that prolonging the treatment period of cells in culture can reduce the effective concentration of EGCG [68]. The environment for cells in culture is also very different from that in tumors. Therefore, we cannot rule out a mechanism just because the *in vitro* effective concentrations of EGCG are higher than we observed *in vivo*. However, it is reasonable to assume that activities effected by low concentrations of EGCG are likely to be more relevant than activities that are produced only at higher concentrations. The results from very high concentrations of catechins in cell culture systems may not be relevant to cancer prevention. Similarly, events due to ROS generated by EGCG extracellularly (which can be prevented by the inclusion of SOD and catalase in the incubation mixture) may not occur *in vivo* [7].

7. CONCLUDING REMARKS

A frequently asked question is that whether all the above reviewed actions of catechins are relevant for cancer prevention or cancer therapy *in vivo*. Apparently, mechanisms suggested by cancer prevention studies in animal models are likely to be relevant. These include the induction of apoptosis in different animal models, inhibition of the phosphorylation of c-Jun and ERK1/2 in lung tumorigenesis models, suppression of phospho-AKT and nuclear β-catenin levels in colon cancer models, inhibition of the IGF/IGF1R axis in colon and prostate cancer models, and suppression of VEGF-dependent angiogenesis in lung and prostate cancer models [13,16,17,19,79]. It is still unclear whether these molecules are direct targets for EGCG or downstream events of the primary action. It is reasonable to assume that some of the high-affinity binding proteins as discussed in Section 4.2 could serve as initial targets, but this point remains to be investigated in animal models.

Because of the broad cancer prevention activities of tea catechins in different animal models, multiple mechanisms are likely to be involved. Even in the same experimental system, one tea catechin such as EGCG may exhibit cancer inhibitory activities via more than one mechanism. The possibility that these mechanisms may work synergistically to exert the cancer preventive activity is interesting and needs to be substantiated. Because of these reasons, precise information about the mechanisms of cancer prevention by tea in humans is even more difficult to obtain. From the limited human studies that are available, action of tea constituents in reducing oxidative stress and enhancing the elimination of carcinogens [35,36] may be important.

As for therapeutic applications, even though laboratory studies with cancer cell lines generated exciting results on the effects of EGCG on cancer signaling, there are only a few clinical studies which have demonstrated the usefulness of EGCG. Although many reports indicate that EGCG acts on specific targets, liver toxicity with high doses of EGCG or catechin mixtures has been well documented [30–32]. The use of lower doses of EGCG in adjuvant therapy together with other agents, such as suggested by Kumazoe *et al.* [90], may be more promising.

ACKNOWLEDGMENTS

The authors thank Ms. Dorothy Wong for her assistance in the preparation of this manuscript. Our research on tea and cancer was supported by U.S. NIH grant CA120915, CA122474, and CA133021.

REFERENCES

[1] C.S. Yang, H. Wang, G.X. Li, Z. Yang, F. Guan, H. Jin, Cancer prevention by tea: Evidence from laboratory studies, Pharmacol. Res. 64 (2011) 113–122.
[2] C.S. Yang, X. Wang, G. Lu, S.C. Picinich, Cancer prevention by tea: Animal studies, molecular mechanisms and human relevance, Nat. Rev. Cancer 9 (2009) 429–439.
[3] J.V. Higdon, B. Frei, Tea catechins and polyphenols: Health effects, metabolism, and antioxidant functions, Crit. Rev. Food Sci. Nutr. 43 (2003) 89–143.
[4] V. Crespy, G. Williamson, A review of the health effects of green tea catechins in in vivo animal models, J. Nutr. 134 (2004) 3431S–3440S.
[5] S. Sang, J.D. Lambert, C.T. Ho, C.S. Yang, The chemistry and biotransformation of tea constituents, Pharmacol. Res. 64 (2011) 87–99.
[6] C.S. Yang, H. Wang, Mechanistic issues concerning cancer prevention by tea catechins, Mol. Nutr. Food Res. 55 (2011) 819–831.
[7] Z. Hou, S. Sang, H. You, M.J. Lee, J. Hong, K.V. Chin, et al., Mechanism of action of (-)-epigallocatechin-3-gallate: Auto-oxidation-dependent inactivation of epidermal growth factor receptor and direct effects on growth inhibition in human esophageal cancer KYSE 150 cells, Cancer Res. 65 (2005) 8049–8056.
[8] C.A. Lipinski, F. Lombardo, B.W. Dominy, P.J. Feeney, Experimental and computational approaches to estimate solubility and permeability in drug discovery and development settings, Adv. Drug Deliv. Rev. 46 (2001) 3–26.
[9] C.S. Yang, S. Sang, J.D. Lambert, M.J. Lee, Bioavailability issues in studying the health effects of plant polyphenolic compounds, Mol. Nutr. Food Res. 52 (Suppl. 1) (2008) S139–S151.
[10] M. Roth, B.N. Timmermann, B. Hagenbuch, Interactions of green tea catechins with organic anion-transporting polypeptides, Drug Metab. Dispos. 39 (2011) 920–926.
[11] K. Jemnitz, K. Heredi-Szabo, J. Janossy, E. Ioja, L. Vereczkey, P. Krajcsi, ABCC2/Abcc2: A multispecific transporter with dominant excretory functions, Drug Metab. Rev. 42 (2010) 402–436.
[12] C. Li, M.J. Lee, S. Sheng, X. Meng, S. Prabhu, B. Winnik, et al., Structural identification of two metabolites of catechins and their kinetics in human urine and blood after tea ingestion, Chem. Res. Toxicol. 13 (2000) 177–184.
[13] J. Ju, J. Hong, J.N. Zhou, Z. Pan, M. Bose, J. Liao, et al., Inhibition of intestinal tumorigenesis in Apcmin/+mice by (-)-epigallocatechin-3-gallate, the major catechin in green tea, Cancer Res. 65 (2005) 10623–10631.
[14] J. Ju, Y. Liu, J. Hong, M.T. Huang, A.H. Conney, C.S. Yang, Effects of green tea and high-fat diet on arachidonic acid metabolism and aberrant crypt foci formation in an azoxymethane-induced colon carcinogenesis mouse model, Nutr. Cancer 46 (2003) 172–178.
[15] H. Xiao, X. Hao, B. Simi, J. Ju, H. Jiang, B.S. Reddy, et al., Green tea polyphenols inhibit colorectal aberrant crypt foci (ACF) formation and prevent oncogenic changes in dysplastic ACF in azoxymethane-treated F344 rats, Carcinogenesis 29 (2008) 113–119.
[16] M. Shimizu, Y. Shirakami, H. Sakai, S. Adachi, K. Hata, Y. Hirose, et al., (-)-Epigallocatechin gallate suppresses azoxymethane-induced colonic premalignant lesions in male C57BL/KsJ-db/db mice, Cancer Prev. Res. (Phila.) 1 (2008) 298–304.

[17] G. Lu, J. Liao, G. Yang, K.R. Reuhl, X. Hao, C.S. Yang, Inhibition of adenoma progression to adenocarcinoma in a 4-(methylnitrosamino)-1-(3-pyridyl)-1-butanone-induced lung tumorigenesis model in A/J mice by tea polyphenols and caffeine, Cancer Res. 66 (2006) 11494–11501.
[18] S. Gupta, K. Hastak, N. Ahmad, J.S. Lewin, H. Mukhtar, Inhibition of prostate carcinogenesis in TRAMP mice by oral infusion of green tea polyphenols, Proc. Natl. Acad. Sci. U.S.A. 98 (2001) 10350–10355.
[19] V.M. Adhami, I.A. Siddiqui, N. Ahmad, S. Gupta, H. Mukhtar, Oral consumption of green tea polyphenols inhibits insulin-like growth factor-I-induced signaling in an autochthonous mouse model of prostate cancer, Cancer Res. 64 (2004) 8715–8722.
[20] T.D. Shanafelt, T.G. Call, C.S. Zent, J.F. Leis, B. LaPlant, D.A. Bowen, et al., Phase 2 trial of daily, oral polyphenon E in patients with asymptomatic, Rai stage 0 to II chronic lymphocytic leukemia, Cancer 119 (2013) 363–370.
[21] B.C. Dickinson, C.J. Chang, Chemistry and biology of reactive oxygen species in signaling or stress responses, Nat. Chem. Biol. 7 (2011) 504–511.
[22] S.P. Hussain, L.J. Hofseth, C.C. Harris, Radical causes of cancer, Nat. Rev. Cancer 3 (2003) 276–285.
[23] G.M. DeNicola, F.A. Karreth, T.J. Humpton, A. Gopinathan, C. Wei, K. Frese, et al., Oncogene-induced Nrf2 transcription promotes ROS detoxification and tumorigenesis, Nature 475 (2011) 106–109.
[24] J. Watson, Oxidants, antioxidants and the current incurability of metastatic cancers, Open Biol. 3 (2013) 120144.
[25] R. Srividhya, V. Jyothilakshmi, K. Arulmathi, V. Senthilkumaran, P. Kalaiselvi, Attenuation of senescence-induced oxidative exacerbations in aged rat brain by (-)-epigallocatechin-3-gallate, Int. J. Dev. Neurosci. 26 (2008) 217–223.
[26] Y. Xu, C.T. Ho, S.G. Amin, C. Han, F.L. Chung, Inhibition of tobacco-specific nitrosamine-induced lung tumorigenesis in A/J mice by green tea and its major polyphenol as antioxidants, Cancer Res. 52 (1992) 3875–3879.
[27] J.L. Schwartz, V. Baker, E. Larios, F.L. Chung, Molecular and cellular effects of green tea on oral cells of smokers: A pilot study, Mol. Nutr. Food Res. 49 (2005) 43–51.
[28] G.Y. Yang, J. Liao, K. Kim, E.J. Yurkow, C.S. Yang, Inhibition of growth and induction of apoptosis in human cancer cell lines by tea polyphenols, Carcinogenesis 19 (1998) 611–616.
[29] G.X. Li, Y.K. Chen, Z. Hou, H. Xiao, H. Jin, G. Lu, et al., Pro-oxidative activities and dose-response relationship of (-)-epigallocatechin-3-gallate in the inhibition of lung cancer cell growth: A comparative study in vivo and in vitro, Carcinogenesis 31 (2010) 902–910.
[30] J.D. Lambert, M.J. Kennett, S. Sang, K.R. Reuhl, J. Ju, C.S. Yang, Hepatotoxicity of high oral dose (-)-epigallocatechin-3-gallate in mice, Food Chem. Toxicol. 48 (2010) 409–416.
[31] J.D. Lambert, S. Sang, C.S. Yang, Possible controversy over dietary polyphenols: Benefits vs risks, Chem. Res. Toxicol. 20 (2007) 583–585.
[32] G. Mazzanti, F. Menniti-Ippolito, P.A. Moro, F. Cassetti, R. Raschetti, C. Santuccio, et al., Hepatotoxicity from green tea: A review of the literature and two unpublished cases, Eur. J. Clin. Pharmacol. 65 (2009) 331–341.
[33] J.D. Hayes, A.T. Dinkova-Kostova, The Nrf2 regulatory network provides an interface between redox and intermediary metabolism, Trends Biochem. Sci. 39 (2014) 199–218.
[34] G. Shen, C. Xu, R. Hu, M.R. Jain, S. Nair, W. Lin, et al., Comparison of (-)-epigallocatechin-3-gallate elicited liver and small intestine gene expression profiles between C57BL/6J mice and C57BL/6J/Nrf2 (-/-) mice, Pharm. Res. 22 (2005) 1805–1820.

[35] H.H. Chow, I.A. Hakim, D.R. Vining, J.A. Crowell, M.E. Tome, J. Ranger-Moore, et al., Modulation of human glutathione s-transferases by polyphenon E intervention, Cancer Epidemiol. Biomarkers Prev. 16 (2007) 1662–1666.
[36] L. Tang, M. Tang, L. Xu, H. Luo, T. Huang, J. Yu, et al., Modulation of aflatoxin biomarkers in human blood and urine by green tea polyphenols intervention, Carcinogenesis 29 (2008) 411–417.
[37] M. Li, A.E. Hagerman, Role of the flavan-3-ol and galloyl moieties in the interaction of (-)-epigallocatechin gallate with serum albumin, J. Agric. Food Chem. 62 (2014) 3768–3775.
[38] M.Z. Fang, Y. Wang, N. Ai, Z. Hou, Y. Sun, H. Lu, et al., Tea polyphenol (-)-epigallocatechin-3-gallate inhibits DNA methyltransferase and reactivates methylation-silenced genes in cancer cell lines, Cancer Res. 63 (2003) 7563–7570.
[39] M. Leone, D. Zhai, S. Sareth, S. Kitada, J.C. Reed, M. Pellecchia, Cancer prevention by tea polyphenols is linked to their direct inhibition of antiapoptotic Bcl-2-family proteins, Cancer Res. 63 (2003) 8118–8121.
[40] S. Ermakova, B.Y. Choi, H.S. Choi, B.S. Kang, A.M. Bode, Z. Dong, The intermediate filament protein vimentin is a new target for epigallocatechin gallate, J. Biol. Chem. 280 (2005) 16882–16890.
[41] M. Li, Z. He, S. Ermakova, D. Zheng, F. Tang, Y.Y. Cho, et al., Direct inhibition of insulin-like growth factor-I receptor kinase activity by (-)-epigallocatechin-3-gallate regulates cell transformation, Cancer Epidemiol. Biomarkers Prev. 16 (2007) 598–605.
[42] Z. He, F. Tang, S. Ermakova, M. Li, Q. Zhao, Y.Y. Cho, et al., Fyn is a novel target of (-)-epigallocatechin gallate in the inhibition of JB6 Cl41 cell transformation, Mol. Carcinog. 47 (2008) 172–183.
[43] S.P. Ermakova, B.S. Kang, B.Y. Choi, H.S. Choi, T.F. Schuster, W.Y. Ma, et al., (-)-Epigallocatechin gallate overcomes resistance to etoposide-induced cell death by targeting the molecular chaperone glucose-regulated protein 78, Cancer Res. 66 (2006) 9260–9269.
[44] J.H. Shim, H.S. Choi, A. Pugliese, S.Y. Lee, J.I. Chae, B.Y. Choi, et al., (-)-Epigallocatechin gallate regulates CD3-mediated T cell receptor signaling in leukemia through the inhibition of ZAP-70 kinase, J. Biol. Chem. 283 (2008) 28370–28379.
[45] J.H. Shim, Z.Y. Su, J.I. Chae, D.J. Kim, F. Zhu, W.Y. Ma, et al., Epigallocatechin gallate suppresses lung cancer cell growth through Ras-GTPase-activating protein SH3 domain-binding protein 1, Cancer Prev. Res. (Phila.) 3 (2010) 670–679.
[46] D.V. Urusova, J.H. Shim, D.J. Kim, S.K. Jung, T.A. Zykova, A. Carper, et al., Epigallocatechin-gallate suppresses tumorigenesis by directly targeting Pin1, Cancer Prev. Res. (Phila.) 4 (2011) 1366–1377.
[47] M. Menegazzi, S. Mariotto, M. Dal Bosco, E. Darra, N. Vaiana, K. Shoji, et al., Direct interaction of natural and synthetic catechins with signal transducer activator of transcription 1 affects both its phosphorylation and activity, FEBS J. 281 (2014) 724–738.
[48] H.Y. Lin, S.C. Hou, S.C. Chen, M.C. Kao, C.C. Yu, S. Funayama, et al., (-)-Epigallocatechin gallate induces Fas/CD95-mediated apoptosis through inhibiting constitutive and IL-6-induced JAK/STAT3 signaling in head and neck squamous cell carcinoma cells, J. Agric. Food Chem. 60 (2012) 2480–2489.
[49] H. Tachibana, K. Koga, Y. Fujimura, K. Yamada, A receptor for green tea polyphenol EGCG, Nat. Struct. Mol. Biol. 11 (2004) 380–381.
[50] Z. Dong, W. Ma, C. Huang, C.S. Yang, Inhibition of tumor promoter-induced activator protein 1 activation and cell transformation by tea polyphenols, (-)-epigallocatechin gallate, and theaflavins, Cancer Res. 57 (1997) 4414–4419.
[51] J.Y. Chung, C. Huang, X. Meng, Z. Dong, C.S. Yang, Inhibition of activator protein 1 activity and cell growth by purified green tea and black tea polyphenols in

H-ras-transformed cells: Structure-activity relationship and mechanisms involved, Cancer Res. 59 (1999) 4610–4617.
[52] J.Y. Chung, J.O. Park, H. Phyu, Z. Dong, C.S. Yang, Mechanisms of inhibition of the Ras-MAP kinase signaling pathway in 30.7b Ras 12 cells by tea polyphenols (-)-epigallocatechin-3-gallate and theaflavin-3,3′-digallate, FASEB J. 15 (2001) 2022–2024.
[53] S. Liu, H. Li, L. Chen, L. Yang, L. Li, Y. Tao, et al., (-)-Epigallocatechin-3-gallate inhibition of Epstein-Barr virus spontaneous lytic infection involves ERK1/2 and PI3-K/Akt signaling in EBV-positive cells, Carcinogenesis 34 (2013) 627–637.
[54] S. Nam, D.M. Smith, Q.P. Dou, Ester bond-containing tea polyphenols potently inhibit proteasome activity in vitro and in vivo, J. Biol. Chem. 276 (2001) 13322–13330.
[55] S. Garbisa, S. Biggin, N. Cavallarin, L. Sartor, R. Benelli, A. Albini, Tumor invasion: Molecular shears blunted by green tea, Nat. Med. 5 (1999) 1216.
[56] S. Garbisa, L. Sartor, S. Biggin, B. Salvato, R. Benelli, A. Albini, Tumor gelatinases and invasion inhibited by the green tea flavanol epigallocatechin-3-gallate, Cancer 91 (2001) 822–832.
[57] G. Deb, V.S. Thakur, A.M. Limaye, S. Gupta, Epigenetic induction of tissue inhibitor of matrix metalloproteinase-3 by green tea polyphenols in breast cancer cells. Mol. Carcinog. (2014). http://dx.doi.org/10.1002/mc.22121.
[58] S. Taniguchi, H. Fujiki, H. Kobayashi, H. Go, K. Miyado, H. Sadano, et al., Effect of (-)-epigallocatechin gallate, the main constituent of green tea, on lung metastasis with mouse B16 melanoma cell lines, Cancer Lett. 65 (1992) 51–54.
[59] M. Pandey, S. Shukla, S. Gupta, Promoter demethylation and chromatin remodeling by green tea polyphenols leads to re-expression of GSTP1 in human prostate cancer cells, Int. J. Cancer 126 (2010) 2520–2533.
[60] E. Navarro-Peran, J. Cabezas-Herrera, F. Garcia-Canovas, M.C. Durrant, R.N. Thorneley, J.N. Rodriguez-Lopez, The antifolate activity of tea catechins, Cancer Res. 65 (2005) 2059–2064.
[61] E.S. Shin, J. Park, J.M. Shin, D. Cho, S.Y. Cho, D.W. Shin, et al., Catechin gallates are NADP+-competitive inhibitors of glucose-6-phosphate dehydrogenase and other enzymes that employ NADP+as a coenzyme, Bioorg. Med. Chem. 16 (2008) 3580–3586.
[62] T. Ishii, T. Mori, T. Tanaka, D. Mizuno, R. Yamaji, S. Kumazawa, et al., Covalent modification of proteins by green tea polyphenol (-)-epigallocatechin-3-gallate through autoxidation, Free Radic. Biol. Med. 45 (2008) 1384–1394.
[63] W. Huang, L. Ding, Q. Huang, H. Hu, S. Liu, X. Yang, et al., Carbonyl reductase 1 as a novel target of (-)-epigallocatechin gallate against hepatocellular carcinoma, Hepatology 52 (2010) 703–714.
[64] C.A. Larsen, R.H. Dashwood, W.H. Bisson, Tea catechins as inhibitors of receptor tyrosine kinases: Mechanistic insights and human relevance, Pharmacol. Res. 62 (2010) 457–464.
[65] M. Shimizu, S. Adachi, M. Masuda, O. Kozawa, H. Moriwaki, Cancer chemoprevention with green tea catechins by targeting receptor tyrosine kinases, Mol. Nutr. Food Res. 55 (2011) 832–843.
[66] H.W. Lo, M.C. Hung, Nuclear EGFR signalling network in cancers: Linking EGFR pathway to cell cycle progression, nitric oxide pathway and patient survival, Br. J. Cancer 94 (2006) 184–188.
[67] Y.C. Liang, S.Y. Lin-shiau, C.F. Chen, J.K. Lin, Suppression of extracellular signals and cell proliferation through EGF receptor binding by (-)-epigallocatechin gallate in human A431 epidermoid carcinoma cells, J. Cell. Biochem. 67 (1997) 55–65.

[68] M. Shimizu, A. Deguchi, J.T. Lim, H. Moriwaki, L. Kopelovich, I.B. Weinstein, (-)-Epigallocatechin gallate and polyphenon E inhibit growth and activation of the epidermal growth factor receptor and human epidermal growth factor receptor-2 signaling pathways in human colon cancer cells, Clin. Cancer Res. 11 (2005) 2735–2746.
[69] S. Adachi, T. Nagao, H.I. Ingolfsson, F.R. Maxfield, O.S. Andersen, L. Kopelovich, et al., The inhibitory effect of (-)-epigallocatechin gallate on activation of the epidermal growth factor receptor is associated with altered lipid order in HT29 colon cancer cells, Cancer Res. 67 (2007) 6493–6501.
[70] S. Adachi, T. Nagao, S. To, A.K. Joe, M. Shimizu, R. Matsushima-Nishiwaki, et al., (-)-Epigallocatechin gallate causes internalization of the epidermal growth factor receptor in human colon cancer cells, Carcinogenesis 29 (2008) 1986–1993.
[71] X. Zhang, H. Zhang, M. Tighiouart, J.E. Lee, H.J. Shin, F.R. Khuri, et al., Synergistic inhibition of head and neck tumor growth by green tea (-)-epigallocatechin-3-gallate and EGFR tyrosine kinase inhibitor, Int. J. Cancer 123 (2008) 1005–1014.
[72] C.S. Mitsiades, N.S. Mitsiades, C.J. McMullan, V. Poulaki, R. Shringarpure, M. Akiyama, et al., Inhibition of the insulin-like growth factor receptor-1 tyrosine kinase activity as a therapeutic strategy for multiple myeloma, other hematologic malignancies, and solid tumors, Cancer Cell 5 (2004) 221–230.
[73] M. Shimizu, A. Deguchi, Y. Hara, H. Moriwaki, I.B. Weinstein, EGCG inhibits activation of the insulin-like growth factor-1 receptor in human colon cancer cells, Biochem. Biophys. Res. Commun. 334 (2005) 947–953.
[74] M. Shimizu, Y. Shirakami, H. Sakai, H. Tatebe, T. Nakagawa, Y. Hara, et al., EGCG inhibits activation of the insulin-like growth factor (IGF)/IGF-1 receptor axis in human hepatocellular carcinoma cells, Cancer Lett. 262 (2008) 10–18.
[75] R. Vittal, Z.E. Selvanayagam, Y. Sun, J. Hong, F. Liu, K.V. Chin, et al., Gene expression changes induced by green tea polyphenol (-)-epigallocatechin-3-gallate in human bronchial epithelial 21BES cells analyzed by DNA microarray, Mol. Cancer Ther. 3 (2004) 1091–1099.
[76] M. Masuda, M. Suzui, J.T. Lim, A. Deguchi, J.W. Soh, I.B. Weinstein, Epigallocatechin-3-gallate decreases VEGF production in head and neck and breast carcinoma cells by inhibiting EGFR-related pathways of signal transduction, J. Exp. Ther. Oncol. 2 (2002) 350–359.
[77] S.K. Rodriguez, W. Guo, L. Liu, M.A. Band, E.K. Paulson, M. Meydani, Green tea catechin, epigallocatechin-3-gallate, inhibits vascular endothelial growth factor angiogenic signaling by disrupting the formation of a receptor complex, Int. J. Cancer 118 (2006) 1635–1644.
[78] M. Shimizu, Y. Shirakami, H. Sakai, Y. Yasuda, M. Kubota, S. Adachi, et al., (-)-Epigallocatechin gallate inhibits growth and activation of the VEGF/VEGFR axis in human colorectal cancer cells, Chem. Biol. Interact. 185 (2010) 247–252.
[79] J. Liao, G.Y. Yang, E.S. Park, X. Meng, Y. Sun, D. Jia, et al., Inhibition of lung carcinogenesis and effects on angiogenesis and apoptosis in A/J mice by oral administration of green tea, Nutr. Cancer 48 (2004) 44–53.
[80] B.H. Zhu, W.H. Zhan, Z.R. Li, Z. Wang, Y.L. He, J.S. Peng, et al., (-)-Epigallocatechin-3-gallate inhibits growth of gastric cancer by reducing VEGF production and angiogenesis, World J. Gastroenterol. 13 (2007) 1162–1169.
[81] P. Peschard, M. Park, From Tpr-Met to Met, tumorigenesis and tubes, Oncogene 26 (2007) 1276–1285.
[82] C. Birchmeier, W. Birchmeier, E. Gherardi, G.F. Vande Woude, Met, metastasis, motility and more, Nat. Rev. Mol. Cell Biol. 4 (2003) 915–925.

[83] R.L. Bigelow, J.A. Cardelli, The green tea catechins, (-)-epigallocatechin-3-gallate (EGCG) and (-)-epicatechin-3-gallate (ECG), inhibit HGF/Met signaling in immortalized and tumorigenic breast epithelial cells, Oncogene 25 (2006) 1922–1930.
[84] C.A. Larsen, W.H. Bisson, R.H. Dashwood, Tea catechins inhibit hepatocyte growth factor receptor (MET kinase) activity in human colon cancer cells: Kinetic and molecular docking studies, J. Med. Chem. 52 (2009) 6543–6545.
[85] Y.C. Lim, H.Y. Park, H.S. Hwang, S.U. Kang, J.H. Pyun, M.H. Lee, et al., (-)-Epigallocatechin-3-gallate (EGCG) inhibits HGF-induced invasion and metastasis in hypopharyngeal carcinoma cells, Cancer Lett. 271 (2008) 140–152.
[86] S.A. Milligan, P. Burke, D.T. Coleman, R.L. Bigelow, J.J. Steffan, J.L. Carroll, et al., The green tea polyphenol EGCG potentiates the antiproliferative activity of c-Met and epidermal growth factor receptor inhibitors in non-small cell lung cancer cells, Clin. Cancer Res. 15 (2009) 4885–4894.
[87] R. Chokor, S. Lamy, B. Annabi, Transcriptional targeting of sphingosine-1-phosphate receptor S1P2 by epigallocatechin-3-gallate prevents sphingosine-1-phosphate-mediated signaling in macrophage-differentiated HL-60 promyelomonocytic leukemia cells, Onco Targets Ther. 7 (2014) 667–677.
[88] D. Umeda, S. Yano, K. Yamada, H. Tachibana, Green tea polyphenol epigallocatechin-3-gallate signaling pathway through 67-kDa laminin receptor, J. Biol. Chem. 283 (2008) 3050–3058.
[89] E. Hong Byun, Y. Fujimura, K. Yamada, H. Tachibana, TLR4 signaling inhibitory pathway induced by green tea polyphenol epigallocatechin-3-gallate through 67-kDa laminin receptor, J. Immunol. 185 (2010) 33–45.
[90] M. Kumazoe, K. Sugihara, S. Tsukamoto, Y. Huang, Y. Tsurudome, T. Suzuki, et al., 67-kDa laminin receptor increases cGMP to induce cancer-selective apoptosis, J. Clin. Invest. 123 (2013) 787–799.
[91] T. Singh, S.K. Katiyar, Green tea polyphenol, (-)-epigallocatechin-3-gallate, induces toxicity in human skin cancer cells by targeting beta-catenin signaling, Toxicol. Appl. Pharmacol. 273 (2013) 418–424.
[92] J. Godeke, S. Maier, M. Eichenmuller, J. Muller-Hocker, D. von Schweinitz, R. Kappler, Epigallocatechin-3-gallate inhibits hepatoblastoma growth by reactivating the Wnt inhibitor SFRP1, Nutr. Cancer 65 (2013) 1200–1207.
[93] V. Nandakumar, M. Vaid, S.K. Katiyar, (-)-Epigallocatechin-3-gallate reactivates silenced tumor suppressor genes, Cip1/p21 and p16INK4a, by reducing DNA methylation and increasing histones acetylation in human skin cancer cells, Carcinogenesis 32 (2011) 537–544.
[94] K.C. Choi, M.G. Jung, Y.H. Lee, J.C. Yoon, S.H. Kwon, H.B. Kang, et al., Epigallocatechin-3-gallate, a histone acetyltransferase inhibitor, inhibits EBV-induced B lymphocyte transformation via suppression of RelA acetylation, Cancer Res. 69 (2009) 583–592.
[95] S.M. Meeran, S.N. Patel, T.H. Chan, T.O. Tollefsbol, A novel prodrug of epigallocatechin-3-gallate: Differential epigenetic hTERT repression in human breast cancer cells, Cancer Prev. Res. (Phila.) 4 (2011) 1243–1254.
[96] D.P. Bartel, MicroRNAs: Target recognition and regulatory functions, Cell 136 (2009) 215–233.
[97] W.P. Tsang, T.T. Kwok, Epigallocatechin gallate up-regulation of miR-16 and induction of apoptosis in human cancer cells, J. Nutr. Biochem. 21 (2010) 140–146.
[98] H. Wang, S. Bian, C.S. Yang, Green tea polyphenol EGCG suppresses lung cancer cell growth through upregulating miR-210 expression caused by stabilizing HIF-1alpha, Carcinogenesis 32 (2011) 1881–1889.

[99] J.Y. Jang, J.K. Lee, Y.K. Jeon, C.W. Kim, Exosome derived from epigallocatechin gallate treated breast cancer cells suppresses tumor growth by inhibiting tumor-associated macrophage infiltration and M2 polarization, BMC Cancer 13 (2013) 421.
[100] M.H. Lee, D.W. Han, S.H. Hyon, J.C. Park, Apoptosis of human fibrosarcoma HT-1080 cells by epigallocatechin-3-O-gallate via induction of p53 and caspases as well as suppression of Bcl-2 and phosphorylated nuclear factor-kappaB, Apoptosis 16 (2011) 75–85.
[101] J.H. Lee, Y.J. Jeong, S.W. Lee, D. Kim, S.J. Oh, H.S. Lim, et al., EGCG induces apoptosis in human laryngeal epidermoid carcinoma Hep2 cells via mitochondria with the release of apoptosis-inducing factor and endonuclease G, Cancer Lett. 290 (2010) 68–75.
[102] V.S. Thakur, K. Gupta, S. Gupta, Green tea polyphenols increase p53 transcriptional activity and acetylation by suppressing class I histone deacetylases, Int. J. Oncol. 41 (2012) 353–361.
[103] C.H. Huang, S.J. Tsai, Y.J. Wang, M.H. Pan, J.Y. Kao, T.D. Way, EGCG inhibits protein synthesis, lipogenesis, and cell cycle progression through activation of AMPK in p53 positive and negative human hepatoma cells, Mol. Nutr. Food Res. 53 (2009) 1156–1165.
[104] D. Duhon, R.L. Bigelow, D.T. Coleman, J.J. Steffan, C. Yu, W. Langston, et al., The polyphenol epigallocatechin-3-gallate affects lipid rafts to block activation of the c-Met receptor in prostate cancer cells, Mol. Carcinog. 49 (2010) 739–749.
[105] Y. Fujimura, K. Yamada, H. Tachibana, A lipid raft-associated 67kDa laminin receptor mediates suppressive effect of epigallocatechin-3-O-gallate on FcepsilonRI expression, Biochem. Biophys. Res. Commun. 336 (2005) 674–681.
[106] T. Kuzuhara, Y. Sei, K. Yamaguchi, M. Suganuma, H. Fujiki, DNA and RNA as new binding targets of green tea catechins, J. Biol. Chem. 281 (2006) 17446–17456.

AUTHOR INDEX

Note: Numbers in Roman indicates the page numbers where the author's work is referred to although the name is not cited in the text. Numbers in Italic indicates to the page numbers on which the complete reference appears.

B

C

D

E

F

K

M

N

Q

R

S

T

Z

SUBJECT INDEX

Note: Page numbers followed by "*f*" indicate figures and "*t*" indicate tables.

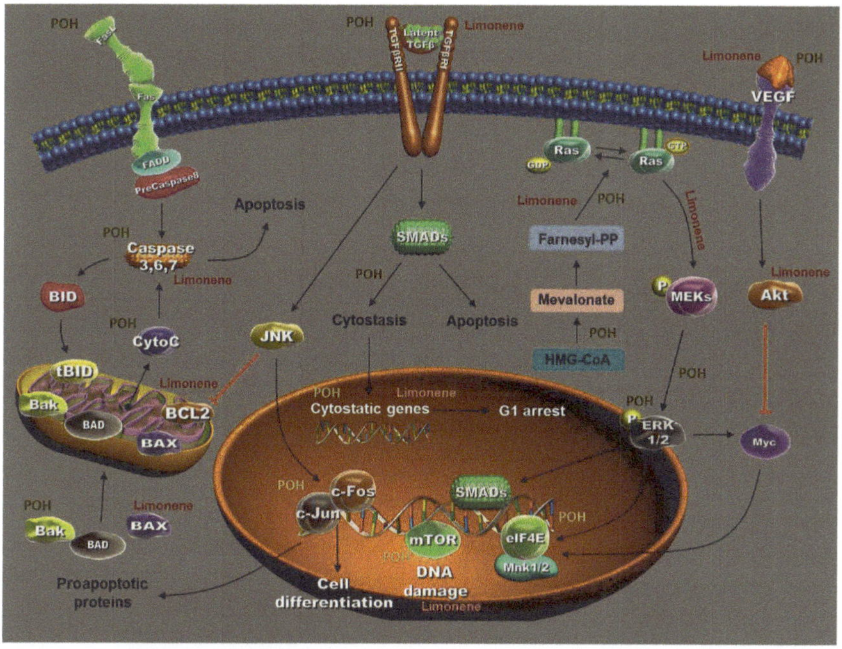

Chapter 2. Figure 2.4 (See legend in text.)

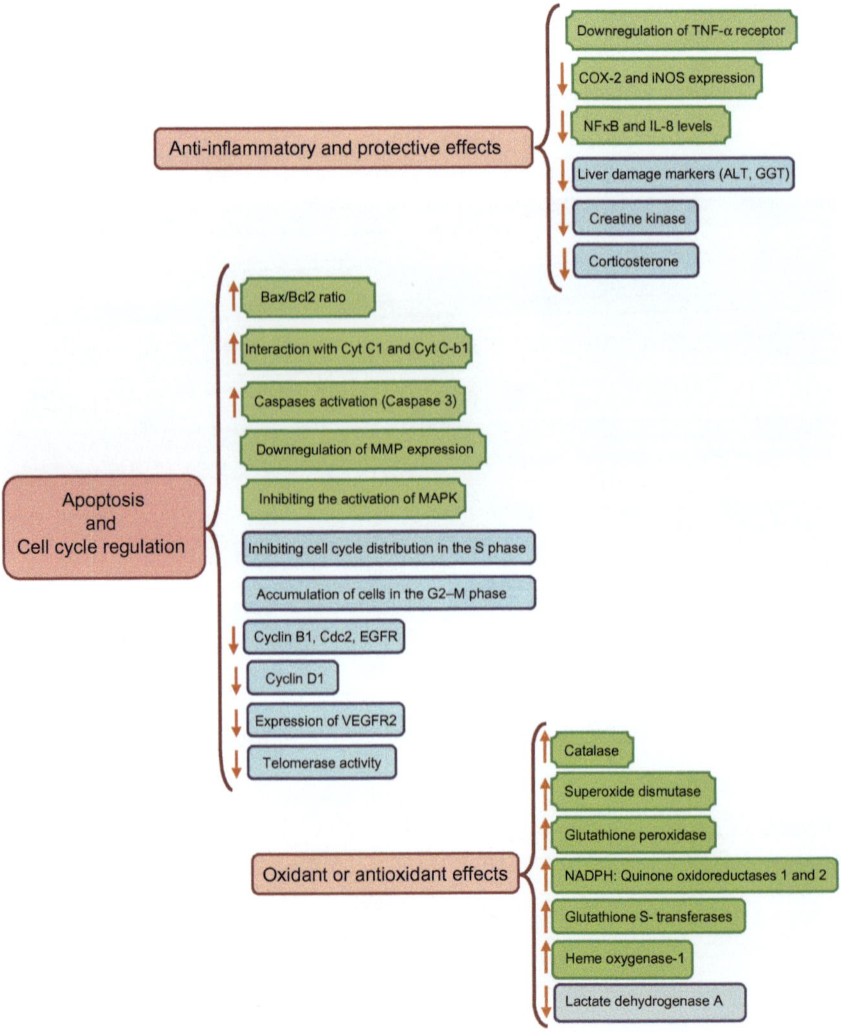

Chapter 4. Figure 4.1 (See legend in text.)

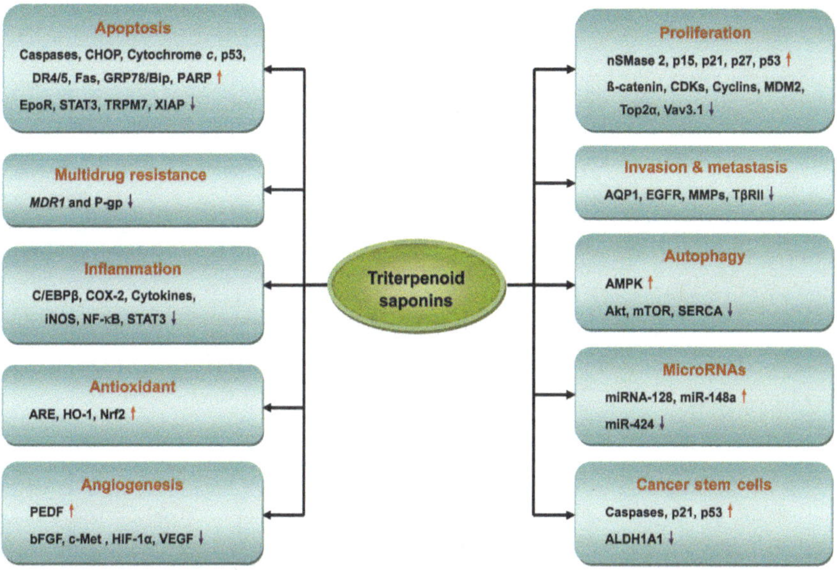

Chapter 6. Figure 6.2 (See legend in text.)

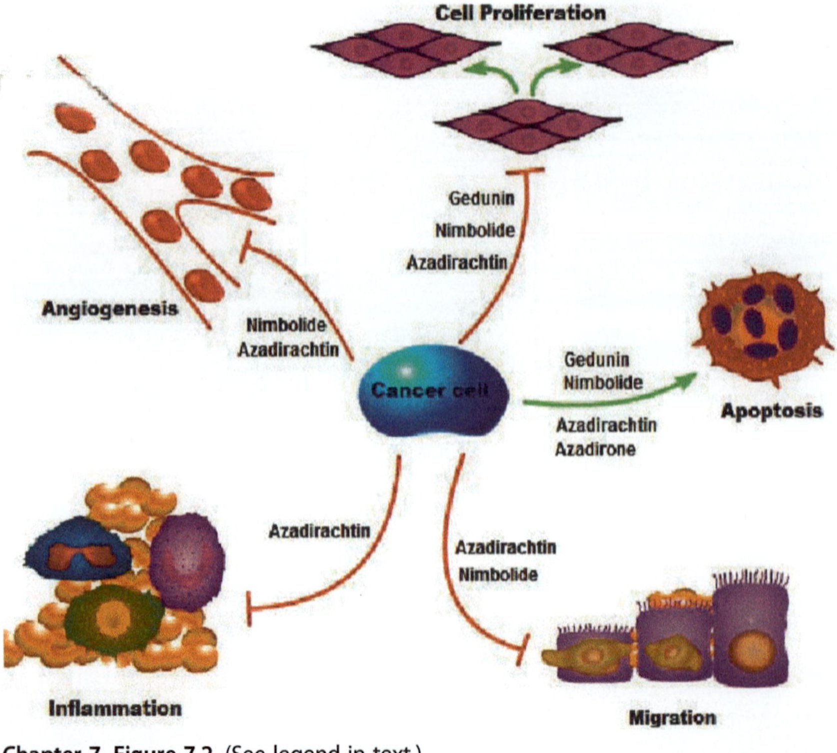

Chapter 7. Figure 7.2 (See legend in text.)

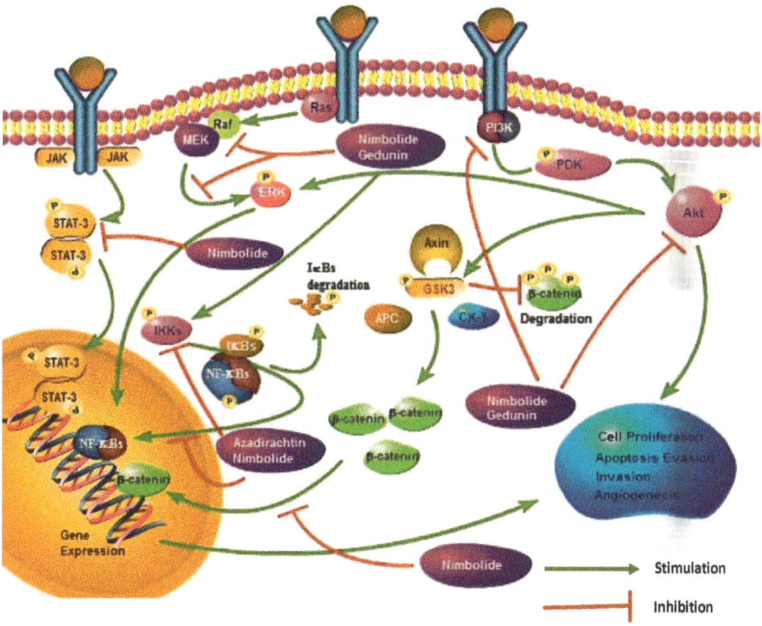

Chapter 7. Figure 7.3 (See legend in text.)

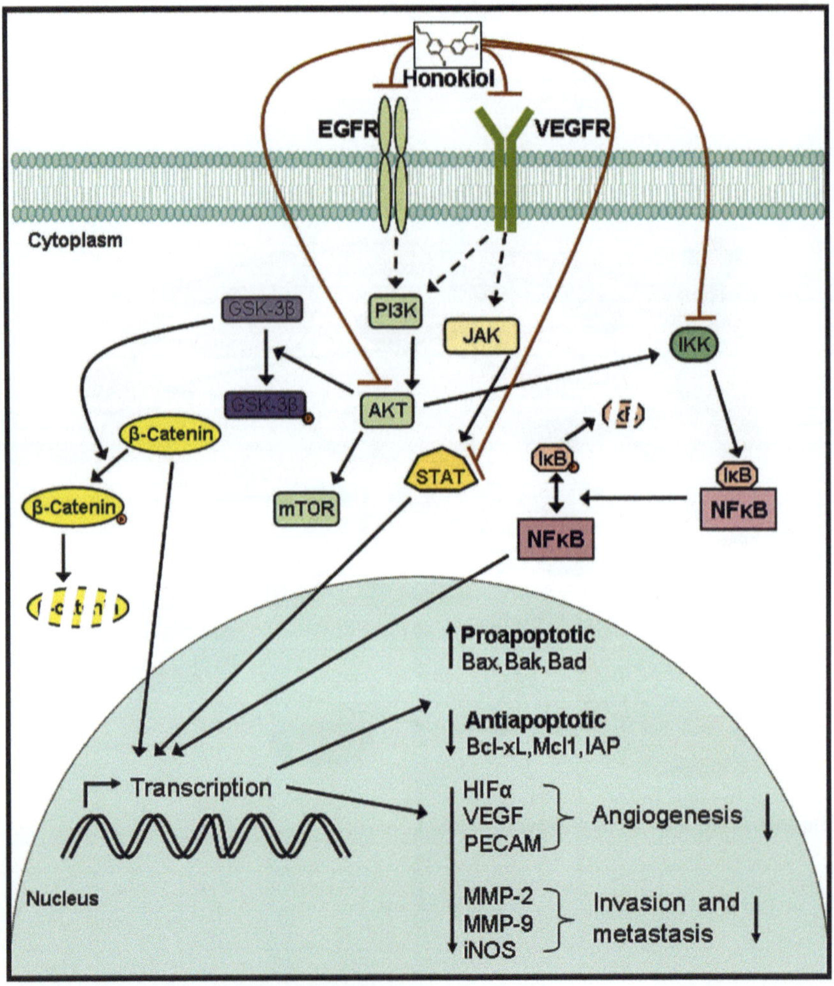

Chapter 9. Figure 9.2 (See legend in text.)

FSC
www.fsc.org
MIX
Papier | Fördert
gute Waldnutzung
FSC® C083411

Kolibri 360 GmbH
Ferdinand-Jühlke-Straße 7
99095 Erfurt, Deutschland
produktsicherheit@kolibri360.de